ROUTLEDGE LIBRARY EDITIONS:
CONSERVATION

Volume 3

THE CONSERVATION OF ECOSYSTEMS AND SPECIES

THE CONSERVATION OF ECOSYSTEMS AND SPECIES

GARETH E. JONES

Routledge
Taylor & Francis Group
LONDON AND NEW YORK

First published in 1987 by Croom Helm Ltd

This edition first published in 2020
by Routledge
2 Park Square, Milton Park, Abingdon, Oxon OX14 4RN

and by Routledge
52 Vanderbilt Avenue, New York, NY 10017

Routledge is an imprint of the Taylor & Francis Group, an informa business

British Library Cataloguing in Publication Data
A catalogue record for this book is available from the British Library

ISBN: 978-0-367-43303-1 (Set)
ISBN: 978-1-00-300237-6 (Set) (ebk)
ISBN: 978-0-367-41675-1 (Volume 3) (hbk)
ISBN: 978-0-367-81566-0 (Volume 3) (ebk)

Publisher's Note
The publisher has gone to great lengths to ensure the quality of this reprint but points out that some imperfections in the original copies may be apparent.

Disclaimer
The publisher has made every effort to trace copyright holders and would welcome correspondence from those they have been unable to trace.

Introduction

It is 32 years since this book was first published and while the scientific reasons for conservation of ecosystem and species have remained largely unaltered, considerable changes in public attitudes and public policy towards conservation have occurred. From the prevailing elitist attitude that existed towards conservation in the 1980s, there has evolved a more pragmatic conservation, one that is recognised as being essential if we are to sustain our lifestyle in at least as good a condition as we now have.

Traditionally, our resources have been classed as either finite or infinite. The latter included the atmosphere and lithosphere – air and water. The rate at which we make use of our resources has increased massively and this has led to new problems relating to over-use of some resources, which in turn has created new pressures on the survival of ecosystems and species.

It is primarily in the area of the infinite resources, sometimes called renewable resources, that new pressures have emerged. It is these infinite resources that have become restricted in their ability to support life in many parts of the world. There are three specific areas of concern regarding the overuse of the biosphere and for which it has become necessary to develop new strategies of conservation management. In particular, the second and third of these have emerged since the late 1980s.

1. The continued growth in human population.
In 1980, the human population was increasing at an annual growth rate of approximately 1.7% to reach an estimated 4.5 billion. In 2020 the human population will be in the order of 7.6 billion, with

a reduced annual growth rate of 0.96%. Statistical projections prepared by the United Nations suggest annual growth rate will fall to 0.1% by the end of the twenty-first century. However, the total population will still increase this century. Statistical modelling suggests a world population of 10.9 billion by 2100.

2. Global warming.
In 1980, climatologists were still doubting that humans could change the climate of our planet. By 2020 we are now certain that the burning of increasing amounts of fossil fuels since the beginning of the Industrial Revolution from about 1750, has resulted in a warming of the atmosphere by about 0.85 °C. Climate scientists can now tell us that even if it were possible to completely and immediately implement the Paris Agreement (UN Climate Change, 2018) and substantially reduce the burning of fossil fuels it would be about 30 years before we saw the maximum effects of global warming and all its consequences on the biosphere.

3. Genetically modified (or edited) organisms (GMOs).
The earlier edition of this book contained only one short paragraph on the potential of genetic engineering as a cause of ecosystem change, while a second paragraph dealt with tropical rainforest as a source of genetic material for pharmaceutical research. Thirty years later, GMOs are now in widespread agricultural use. Genetic modification is a technology that involves inserting DNA into the genome of an organism, Royal Society (2016). For example, to produce a GM plant, new DNA is transferred into plant cells. Usually, the cells are then grown in tissue culture where they develop into plants. The seeds produced by these plants will inherit the new DNA.

Genetic engineering techniques when applied to agricultural plants and animals undoubtedly has major advantages in allowing introduction of new traits that permit resistance to disease, greater tolerance of drought and heat and more reliability of growth performance. Commercial sale of a genetically modified food crop for human consumption began with tomatoes in 1994. Already, genetically modified corn, soybeans, oil seed rape (canola) and cotton are extensively grown worldwide. The first mammal to be cloned, using the process of nuclear transfer was achieved in July

1996 when a lamb (known as 'Dolly the sheep') was born. No genetically modified animals are currently used as a human food source.

Since the mid-1990s the rate of development in gene manipulation technology has proliferated. A commercially available method known as the CRISPR-Cas9 gene editing system has enabled laboratories around the world to produce genetically modified organisms. Despite the assurances from scientists that GMO crops are safe for human consumption, public concern over the use of gene editing techniques remains high, especially in relation to human health. Important as the use of GMO agricultural crops may be for feeding the increasing world population, concern exists that the GMO species will, one day, 'escape' from where they are grown and interbreed with 'wild' members of the species. It is also certain that animals, especially invertebrates, will feed on GMO crops, but we currently have little knowledge of how the influence of GMOs will spread to other species in the biosphere. Will GMO escapees insert themselves into ecosystems and live alongside wild species or will they interbreed or replace them entirely?

Conservation in the last decades of the twentieth century was the responsibility of individual countries and there was little input or consideration of other, adjacent, countries. It was apparent that while individual national conservation strategies were a useful first stage, what was required was a 'One World' strategy to cope with conservation issues that were not confined by national boundaries. For example, the burning of rainforest in Brazil or Indonesia was a national decision, whereas the repercussions of burning were of international significance. By the early 2000s the purpose of forest burning had become large scale and clearly visible from widely available satellite imagery. In the Amazonian rainforest, between May 2000 and August 2006, Brazil lost nearly 150,000 square kilometres (58,000 sq. mi) of rainforest, an area larger than Greece. Not only does a forest fire destroy the biological diversity within the burnt area, the burning process releases vast amounts of particulate matter and greenhouse gases (mainly CO_2) to the atmosphere and has an unpredictable impact on ecosystems throughout the world. Satellite images showed how clouds of smoke from forest fires drifted thousands of kilometres with no respect for international boundaries.

There was increasing evidence by the mid-1990s that some human activities were causing highly damaging impacts on specific ecosystems and species within the biosphere. Tropical rainforests were under particular threat as were mangrove forests, fresh water wetlands, and upland and high latitude peat bogs. Not only was the extent of damage to these habitats of serious concern, but also the rate at which it was occurring. The last decades of the twentieth century saw increasing urbanisation and the rise of high-tech industrialisation that made use of huge amounts of energy which in turn released commensurate amounts of air pollutants. In agriculture, there was increasing intensification of production that often relied on the indiscriminate use of herbicides and pesticides. Intensification of farming onto marginal land also occurred leading to further removal of forest, and often resulted in soil erosion.

If left untreated, it was feared these threats would cause a critical decline in the ability of the biosphere to support life, including that of our own species. A number of scientific organisations began to arrange multi-national and multi-disciplinary conferences, the purpose of which was to examine a wide variety of topics best described as ìenvironmental interestsî. The conferences gradually grew in stature and began to attract leaders from the world of politics, agriculture, land managers and conservation groups. Non-government organisations (NGOs) were also represented. Foremost among the organisations responsible for organising these meetings has been the United Nations, and the most significant of a range of meetings has been the Conference of Parties of The United Nations Framework Convention on Climate Change, commonly known as the COP. It has met annually since 1995, and the first meeting is known as COP1. The 21st Session of the COP (COP21), held in Paris, France, in December 2015, was historic in its outcome as it resulted in the first international climate agreement, in which all 197 world nations agreed to work towards limiting global warming to a maximum of 2 °C by the end of the century.

Unfortunately, on 1st June 2017 the USA removed itself from the agreement, claiming it would put the US economy at a disadvantage. Despite criticisms from some nations that the timescales and the thresholds to be implemented in the COP21 are too lenient there is little doubt that it represents a major diplomatic achievement and has raised the awareness of politicians and the general public of the problem of global warming.

As the role of the U.S. as a leader in conservation strategy declined then so its place was taken by the European Union (EU). An extensive number of policies have been developed and accepted by all EU Member states. These include: Habitats Directive (1992); Natura 2000 (2000); Biodiversity Strategy (2010); and Nature and Biodiversity Law (1992 to present). These have become compulsory parts of the legal infrastructure that protects the biosphere components.

If we can identify the one most important change in the way in which our attitude towards conservation of ecosystems and species has developed over the last 30 years it is the way in which environmental legislation has emerged as the means to achieve international agreement on how we take care of the biosphere.

References

Root, T.L., J.T. Price, K.R. Hall, S.H. Schneider, C. Rozenzweig and J.A. Pounds (2003) *Fingerprints of global warming on wild animals and plants. Nature* 421, pp 57–60.

Royal Society (2016) What is genetic modification (GM) of crops and how is it done? https://royalsociety.org/topics-policy/projects/gm-plants/what-is-gm-and-how-is-it-done/?gclid=EAIaIQobChMIqrurmIO-4wIVRbDtCh0xywJnEAAYASAAEgLfOvD_BwE

United Nations Climate Change (2018) The Paris Agreement. https://unfccc.int/process-and-meetings/the-paris-agreement/the-paris-agreement

The Conservation of Ecosystems and Species

GARETH E. JONES

CROOM HELM
London • New York • Sydney

Croom Helm Ltd, Provident House, Burrell Row,
Beckenham, Kent, BR3 1AT
Croom Helm Australia, 44-50 Waterloo Road,
North Ryde, 2113, New South Wales

Published in the USA by
Croom Helm
in association with Methuen, Inc.
29 West 35th Street
New York, NY 10001

British Library Cataloguing in Publication Data

Jones, G.E.
The conservation of ecosystems and species.
— (The Croom Helm natural environment)
1. Nature conservation
I. Title
639.9 QH75

ISBN 0-7099-1463-6

Library of Congress Cataloging-in-Publication Data

Jones, G.E. (Gareth E.), 1944-
The conservation of ecosystems and species.

(The Croom Helm natural environment — problems and management series)
1. Nature conservation. I. Title. II. Series.
QH75.J66 1987 333.7'2 87-6781
ISBN 0-7099-1463-6

Printed and bound in Great Britain by Mackays of Chatham Ltd, Kent

CONTENTS

TABLES

Tables

FIGURES

Figures

APPENDICES

<u>Appendix</u>

PREFACE

Mankind differs from all other animal species on this planet in a multitude of ways. One of the unique features of our species is the tendency to exploit the biosphere and its species so that we may flourish. We consider that our use of the biosphere is a 'right' - a resource available for man's well-being. Over the many thousands of years in which we have evolved, our ability to use the biosphere has come to know few bounds. We hunt animals, we clear vegetation to make way for agriculture, we build our cities and our lines of communications, we pour out our pollutants into the biosphere all with little or no thought of the effects our actions may have upon the way in which the biosphere functions.

A fundamental expectation of every inhabitant of the free world is to participate in the 'rights' which are bestowed by a democratic civilisation. Free speech, freedom to choose a political leader, the right to live wherever one wishes. But freedom asks one thing in return. Obligation. The obligation of behaving in a socially acceptable manner. This has become one of the characteristics of civilised society; the development of a human 'ethic' - the sense of doing right - is an unique human trait.

Unfortunately, we do not extend the concept of the human ethic to include the care of the biosphere. Resources are greedily used up as if there is no tomorrow. Chemical and nuclear pollutants are released into the biosphere where they may remain in toxic state for thousands of years. We mistreat our biosphere in the full knowledge that what we are doing is wrong.

This book sets out to look at the scientific basis and needs of conservation of species and ecosystems. First, the history of the conservation movement is examined before examining the fundamental nature of the biosphere for <u>all</u> our necessities of life. Chapter Three looks at the development of our species from the earliest beginnings when man was no more than another mammal, to the emergence of modern 'super' man, a <u>Homo colossus</u> when compared with our ancestors. Chapter Four examines the reasons for the development of a conservation ethic and the different ways we evaluate the biosphere. We are still unable to produce a realistic index to quantify the whole biosphere and instead must content ourselves with working with individual parts of the system. At best we can work at the landscape level - the theme of Chapter Five, though more often we can only operate at the level of the species (Chapter Six) and more rarely with whole ecosystems (Chapter Seven). Finally, Chapters Eight and Nine examine the roles of planning, economics and politics upon the development of a successful conservation ethic.

This book makes no claim to be a complete study of conservation. Whole areas have been only touched upon, while other areas have received my personal interpretations, based upon facts and analysed with scientific analysis.

The conservation message is clear. We must develop and apply a conservation ethic if we wish our planet to survive. A conservation ethic can help provide future generations of mankind with the same freedom of choice that we now enjoy.

Many people have helped in various ways with the preparation of this book. I am especially indebted to Miss Christine Meek for typing all of the chapters in an efficient and professional manner. The Department of Geography at Strathclyde University provided the necessary computer facilities on which to prepare the text. Mrs L Nelson typed the bibliography, Mrs E Harvey prepared the art work, Mr S Keenan assisted with word processing and Miss V Wilson, Mr P Friel and Mr J Ferguson helped with the computer drawn diagrams. As always my family have proved very understanding of the long hours I have spent, particularly in the final stages of preparing the book. My final thanks goes to Mr Peter Sowden and Ms Ingrid Curl of Croom Helm and Dr Chris Park, Series Editor, for their patience and helpful advice throughout the preparation of the book.

ACKNOWLEDGEMENTS

The author and publisher gratefully acknowledge permission granted by the following to reprint or modify copyright material:

Scott, Foresman for Table 2.1; Wiley, for Table 2.2; Bonanza Books, for Fig 3.2; C.U.P., for Fig. 3.3; Geographical Magazine, for Fig, 3.6; Applied Geography, for Table 4.3; David & Charles, for Table 5.1; HMSO, for Fig. 5.1; University of Waterloo, for Table 5.3; Longman, for Fig. 5.4; J.P. Harroy, for Table 5.4; C.U.P., for Figs. 6.3 and 7.1; Basic Books Inc., for Table 6.1; Longman, for Fig. 6.7; Ann Arbor Science Publishers Inc. for Tables 8.1 and 8.2,

Chapter One

CONSERVATION AND ECOLOGICAL THEORY

There is all too often an aura of suspicion about conservation and the people who advocate its practice (Moore, 1968). It has been regarded as a minority interest, practised by people whose sole intent appears to be the imposition of limits or restraints upon the personal freedom of others. The commonplace understanding of conservation is associated, for example, with not being allowed to cut down trees, not killing whales, not demolishing old, derelict buildings or not releasing pollutants to the atmosphere.

The repressive function of conservation is, perhaps, the inevitable result of decades of legislation during which politicians have reacted to scientific reports and also to public pressure groups who have recommended that control measures should be imposed on the ways in which we are using, and often destroying, the resources of our planet.

In the context of the total time-span over which man has used the biosphere and its resources, the legislative influence on conservation attitudes is a relatively recent phenomenon. The first 'law' concerning the conservation of animals was that passed by the Assembly of Bermuda sometime prior to the year 1650 when a democratically elected group of people voted to control the number of turtles that could be taken from the sea-shores around Bermuda (reported on page 466, Warren and Goldsmith, 1974).

The advent of industrialisation in north-west Europe from the early 1800s took scant regard of what would now be called 'environmental considerations'. Mineral resources were ripped from the ground with no regard for human safety nor of the adverse effects such exploitive behaviour would have on plants and animals. Local pollution levels reached horrific levels. The account of Barber who toured the South Wales metalliferous industries in 1803 reads as follows:

> The effect in passing through these dismal buildings (the numerous small copper works that scattered the lower Swansea valley at the beginning of the nineteenth century) contrasted by the vivid glare of the furnaces and the liquid fire of the pouring metal is, to a stranger, very striking. I was much surprised by the quantities of condensed sulphurous vapour that yellowed the roof of the building. Sulphur often forms the greatest part of the ore (copper ore); yet no means are employed to collect the vapour which might easily be managed, and could not fail to turn to a source of profit: at the same time it would save the health of the workmen and spare the vegetation, which appear stinted for a considerable distance by the noxious effluvia.
> (Barber, 1803).

In 1863 the British government passed the Alkali Act which established an inspectorate with the powers to visit industries in order to check on health and safety procedures and to recommend to industrialists ways of minimising pollution of all forms (World Conservation Strategy, 1983). This act marked the beginning of a long and increasingly complex battle between government, industrialists and the general public to ensure that adequate standards of public safety and cleanliness of the environment were maintained.

Possibly the most significant piece of legislation in the short history of conservation was the Conservation Act which made its way through the House of Representatives in the United States of America and which entered the statute books in March, 1872. Under this Act land was set aside for the exclusive benefit and enjoyment of the public. The areas were to be called National Parks and the first location to be so called was the area now designated Yellowstone National Park. The significance of this Act cannot be over-emphasised for it was to stimulate an interest in conservation in almost every other nation of the world.

The early attempts at controlling the use of natural resources were usually the result of intense lobbying of parliamentarians by small, but very influential groups of individuals, notably landowners, scientists and clergy (Gilg, 1981). The objectives of these groups were often very one-sided and involved the preservation of minority interests, for example, the prevention of indiscriminate killing of wild animals such as foxes, badgers and otters. This might seem a laudable request but it was often made by wealthy land owners anxious to preserve the hunting value of their land. Gradually, the intentions of the pressure groups became of genuinely good intent although interest was directed mainly at the level of the individual species. For example, the Wild Cat (Meles meles) of the Scottish Highlands was provided with legal protection but its numbers continued to decline partly because no attempt was made to prevent the destruction of its habitat.

It was events in the USA at the beginning of the twentieth century which set the pattern for the development of a modern conservation policy. The North American landmass had experienced a progressive and relatively well organised exploration of its resources, beginning in the 1750s and virtually completed by the early 1900s (Allen, 1814). Unlike so many of the early human migrations, that which occurred in North America was undertaken by an aggressive, technological, innovative group of settlers. Within the relatively short time of one hundred years the indigenous Indian population had been cleared to make way for agriculture; soils had been eroded and hunting had reduced the population sizes of many of the indigenous mammals and birds.

Under these circumstances of rapid environmental change, several North American scientists, notably Marsh (1874), began voicing their concern over the loss of species and, of particular relevance, the disappearance of 'wilderness-type environments'. The significance of the North American concern with wilderness type landscapes is of major relevance in the emergence of a conservation ethic and the issue reappears frequently throughout this book.

The population of North America at the turn of the

century was about 82 millions (figure quoted in Durand, 1971). With such a small population spread thinly over a large landmass environmental conflicts, when and where they did occur, were extremely localised. The inhabitants of the USA had no wish to follow the pattern which had already been set in north-west Europe in which legislation was seen to be the way to control the use of the environment and through which a conservation policy could be implemented (see p. 1). Instead, they favoured an overall approach in which all the resources of an area (the physical landscape resources along with the plants and animals which inhabited that landscape) should be evaluated and to which a 'conservation' policy applied (Price, 1911). Pinchot (1936) has provided an interesting review of the early development of conservation in North America while Beazley (1967) has re-examined the initial attitudes towards conservation and has attempted a re-interpretation of the early conservation movement through the attitudes which prevailed at the end of the 1960s (pre-energy crisis). Beazley (op.cit) has defined the late nineteenth century approach to conservation as "the establishment and observation of economic, social and politically acceptable norms, standards, patterns or models of behaviour in the use of natural resources by a given society".

Unfortunately, definitions, unlike the laws or theorems of pure science can be interpreted in many different ways. Of equal importance is an appreciation of the points which have been omitted from the definition. In Beazley's definition above, ambiguity undoubtedly arises over the interpretation of the term 'acceptable norm'. There is no base line for an acceptable norm. Different societies will view the term from their own internal standpoint; even different individuals within a society will set different levels of acceptable usage upon a natural resource. The acceptable norm will also fluctuate with time, and will depend amongst other things upon the population density, the technological capability of the society, the prevailing social and economic circumstances and, perhaps of greatest significance, the prevailing political viewpoint.

A conservation attitude, or policy, along with a definition of conservation must, therefore, be capable of undergoing modification in order to accommodate the circumstances which prevail at a given point in time. But a conservation attitude is not a passive, subservient whim which reacts to external pressures. A conservation attitude can itself exert pressure upon governments, industrialists, economists and society and by so doing raises our environmental standards.

Motivation for Conservation

Although the early North American writers on conservation stressed the need to retain wilderness landscapes, the underlying motivation was that of economic forces. If a resource was judged to have an economic value, either at the present or possibly in the near future, then that resource was judged worthy of conservation. By contrast, biological considerations were glaringly absent.

It is unlikely that a conservation policy can be successful if based only on economic considerations. The inclusion of biological and ecological factors are of paramount importance. For example, if a population of fish is to be conserved for its economic value, then the inherent reproductive rate of the fish must be known so that the annual fish catch does not exceed the renewal rate. A classic example of economic factors being allowed to over-rule biological factors can be seen from the Peruvian anchovy fishing industry. In this instance the Peruvian government's wish to increase the sale of fish-meal to the beef-lot farmers of USA led to over exploitation of the anchovy fish stock with its ultimate collapse (Paulick, 1971)

Also, it is necessary to know such basic information as the sex-ratio, the proportion of resident and migratory members of the population, the age structure, the primary and secondary productivity of the organic components, the climatic inputs, the soil types, indeed, the complete list is huge, but for a conservation policy to be successful it is necessary to know as much as possible about the ecological and biological factors pertaining to the ecosystem which is to be conserved.

Unfortunately, the predominance of an economic motivation for conservation as outlined in the definition by Beazley (op.cit) has persisted despite our fuller understanding of the requirements necessary for successful conservation. The economic arguments are clearly seen in a passage of a speech to the US Congress by President J.F. Kennedy in 1962:

> Conservation can be defined as the wise use of our natural environment: it is, in the final analysis, the highest form of national thrift - the prevention of waste and despoilment while preserving, improving and renewing the quality and usefulness of all our resources.

At the time, President Kennedy's conservation definition was hailed as a major statement by a politician but careful reading of the statement will reveal major areas of ambiguity. No level is set for the "wise use of our natural environment". There is a strong implication that only those resources with an economic value are worthy of conservation. By implication, it suggests that all the biospheric components which do not have an economic value are worthless and not worthy of inclusion in a conservation policy.

Conservation - A Change of Direction

Towards the end of the 1960s and early 1970s a marked change in approach towards conservation became apparent. This was due to a number of factors.

a) The realisation by industrialists that mineral resources and energy supplies were not inexhaustible and that rising prices of many commodities would encourage the re-cycling of scrap materials.

b) The publication of a number of 'key' books and research projects which highlighted the problems of resource depletion and species extinction - for example, Silent Spring, by Rachel Carson (1962).

c) Major concern over the explosion of the human population and the demands this would place on the future levels of resource use (US Commission on Population Growth, 1972).

d) The appearance on television of frequent documentary programmes on over-population, pollution, energy crises. In Britain, a major series was devoted to the status of mankind;

The Ascent of Man, by Professor Bronowski (1973).

Fraser-Darling (1967) suggested that the new attitudes displayed towards conservation introduced a much greater application of applied ecological theory than had hitherto been the case. For the first time the arguments for conservation were shown to involve the widening of possibilities and not a restriction of choice. The diversity of the biosphere along with its self-sustaining output of biological yield and its capabilities for absorbing pollutants were explained to the non-biological world. Of even greater importance it was shown that man must work in harmony with nature if all the advantages of the biosphere were to be optimised. An opposition to working within the natural biosphere cycles was shown to result in the ultimate destruction of the system with a resultant hardship for mankind.

One of the consequences of the new awareness towards conservation was the realisation that those resources we had conveniently labelled as 'renewable' could be over exploited and could ultimately disappear altogether if the level of use was set high.

Non Renewable Resources

The division of resources into the two categories of 'renewable' and 'non-renewable' had been a convenient nicety first proposed by Ciriacy-Wantrup (1952). In this division of resources the non-renewable resources were primarily the non-living, mineral substances such as coal, iron ore, copper, lead and zinc. Once used, the resource was depleted and eventually exhausted. Only the passage of very considerable lengths of time would enable the natural geological cycling of materials to generate new resources. The original definition of the non-renewable resource overlooked the possibility of recycling of materials. Thus glass, lead, copper and water are all now frequently recycled (Detwyler, 1971; Environmental Resources Ltd., 1986).

Renewable Resources

Renewable resources were envisaged as those commodities which were capable of natural renewal. The term was applied mainly to biological components (animals and plants), and also to the soil, air and water although the latter three components are, of course, not part of the biological world. They were included because their renewal rates were sufficiently rapid to fit within man's own timescale.

It is paradoxical that many of the renewable resources have disappeared, for ever, from our planet. They have been subject to over-use, their habitats have been destroyed or polluted while some have been judged to be competitive for the same resources as needed by man. Species extinction has, and continues, to eliminate many hundreds of species. Allen (1980) has provided figures for the threatened extinction animals, see Table 1.1.

Table 1.1 Threat of Extinction to Animals

Extinction threat	% of all animals under threat*
Destruction of habitat	67
Over exploitation	37
Impact of introduced exotics	19
Competition with man for food	7
Accidental killing	2

* Figures add up to more than 100% because many species are threatened from more than one cause. Figures from Allen (1980) <u>How to Spare the World.</u>

Some Definitions of Conservation

Definitions are like the weather - liable to change! It is easier to explain what is <u>not</u> conservation rather than to provide a comprehensive, fully acceptable definition. As has already been shown on pp. 4 - 6, conservation has often been defined with economic considerations very much to the fore whereas for conservation to be fully and properly implemented it is necessary for biological and ecological dimensions to be added.

There are three terms which are commonly used in a somewhat arbitrary and often interchangeable manner to imply the act of conservation.These are: a) preservation, b) protectionism (or control) and c) conservation. Of these, the first and second can, at least, be considered evolutionary steps which may form the precursors of conservation proper.

Preservation is the prevention of destruction; in the contemporary context, preservation is applied to extremely rare, and hence valuable objects. The traditional preservationist approach can be seen in museums where ancient artefacts are displayed behind glass and in a controlled environment. The preserved object is inevitably an inert (dead or lifeless) object. More recently, preservation has been extended to inanimate objects such as historic buildings, costume, vintage motor cars and - of particular interest - to language, for example the original Swiss language of Romansh. Preservation cannot be mistaken for conservation.

Protectionism implies the defence of an object or commodity which has been overused or misused in the past. In its extreme form, absolute protectionism is synonymous with preservation in that no change from an existing state is allowed - apart from renovation necessary to repair damage. More typically, protectionism limits or restricts the level of use, for example the European Economic Community operates a strict strategy of 'fishing quotas' whereby member states are allowed to take fish from within European coastal limits up to annually set tonnages. These values are based upon scientifically set guide lines (Ehrlich, 1982; FAO, 1983).

Application of a protection policy on a species, habitat or resource is usually made in order to prolong its existence. For a non-renewable resource the protection policy may be dictated mainly by economic factors, for example, in times of surplus the Texas oil fields produce relatively little oil, while when shortages push world prices higher then so the Texas oil industry flourishes.

Renewable resources cannot be managed in such a simple, deterministic manner. For a biotic resource such as a forest, or an animal population, the usable yield of the resource will be a function of the age:sex ratio, biotic potential, migration and fluctuations in the physical environment circumstances (for example the effect of climate, pollution, soil fertility). Any attempt to apply a protectionist policy to living organisms must therefore be based on a sound ecological basis with decision-making based upon up-to-date information of the system dynamics.

The traditional North American approach to conservation which has been widely adopted by other nations is more akin to protectionism than to true conservation. The definitions given by President Kennedy (p. 5) and by Beazley (p. 4) reflect greater concern for economic arguments rather than ecological factors. A protectionist approach can, however, be easily extended to include ecological factors and, as such, moves closer to the concept of true conservation.

Conservation proper embraces all the points that have been made for preservation and some of those that relate to protectionism. It goes further, however, in that a conservation policy involves a conceptual appreciation of the need to use biospheric resources in ways which place the least restriction upon the future well-being of organic and inorganic resources. In its broadest sense, conservation is now involved with husbanding the resources of the whole biosphere. Pettigrew (1982) has suggested that a complete conservation policy involves the incorporation of a 'conservation ethic' into the everyday life styles of human society. Instead of showing a myopic interest in economic growth rates we should be focusing upon the ability of society to be self-sustaining in terms of energy and material use.

Aldo Leopold was the first person to articulate the need for an ethical approach to conservation in his now classic book A Sand County Almanac (Leopold, 1949). In its extreme case, conservation becomes an elitist attitude which can be afforded only by the most wealthy and best educated of nations. Conservation of the type advocated by Leopold brings it into direct conflict with competing interests (mainly those of the politician and industrialist) and for conservation to make progess the case for conservation must be presented in a form which is directly understandable by non-scientific and often anti-conservationist bodies (Sandbach, 1980).

The elitist approach to conservation in which species and their ecosystems are conserved for their own interest and nothing else is an unrealistic attitude in this increasingly overcrowded planet. Conservationists are forced to compromise. Only certain plant and animals species can be conserved in their natural habitats. Decisions must be made about which species and ecosystems are the most worthy of conservation. Making these decisions is difficult; it requires the best scientific information. It also requires the ability to look to the future in order to predict the pressures and requirements that mankind (in the twenty-first century and beyond) will place upon the biosphere.

A successful conservation policy thus becomes something of a juggling act. It must combine the requirements of the biological scientist for absolute conservation along with the needs of the industrialist keen to produce material goods to be sold on the free market. It must compromise the immediate demands set by present day society with the anticipated demands of future societies. It must convince society that non-materialistic values, such as amenity, recreation capacity, rarity values and self-sustainability, are of equal importance to export earnings, full employment and economic growth. The latter group of values can be costed with an acceptable degree of accuracy whereas the conservation values cannot easily be converted into an economic dimension (see Chapter Nine).

To the conservationist, the industrialised society in which the majority of the world's population now live, is an example of an exploitive system based on finite resources. Eventually, the resource base will be exhausted and the industrialised system will 'crash'. By adopting a conservationist approach in which resources are used "for the purpose of maximising the aesthetic, educational, recreational and economic benefits to society" (Green, 1985), the time at which our biosphere will become depleted of essential raw materials is extended into the future.

Any society which based its wealth and health upon an infrastructure which includes a sound ecological base could expect to achieve benefits in a wide variety of ways. For example, the landscape would become more diversified in appearance. This would be the result of the wide range of different habitats encouraged by the unrestricted operation of the environmental variables. Ecosystems would show less evidence of disturbance; plant and animal diversity would be enhanced; erosion would be reduced and this might be reflected in reduced rates of soil erosion from agricultural land.

A conservation policy could also help reduce the indiscriminate overuse of physical resources. Re-use through recycling of scarce raw materials would minimise the need to mine, or import costly raw materials. An infrastructure based upon ecological conservation can be likened to an insurance policy; the beneficiaries of the conservation policy could expect a better quality of life though the quantity of material resources may be reduced and their proportion altered. The conservation policy would also make available more time in which man could come to terms with the advantages of pursuing an even stronger conservation approach to the use of biosphere resources.

No nation currently operates a conservation policy in which ecological factors outweigh the economic factors. Every country is locked into the battle in which current levels of growth and development are seen as the main indicators of success. Warning signs associated with the misuse and overuse of the biosphere are there for all to see: pollution, famine, soil erosion, eutrophication of rivers and lakes, shortages of key resources.

To change from a system orientated towards constant growth and expansionism to one in which values of constancy are judged to be of at least equal importance would be traumatic, expensive and involve fundamental changes to our established life styles. For example, a conservation-based system might impose stringent limits on the amount and concentration of pollutants which industries could release to the biosphere. Additional pollution control would be required on industrial, transport and domestic sources of waste.

Initially, these additional controls would be expensive (involving development costs, installation, and altered production techniques). Industrialists would, no doubt, claim compensation for these increased production costs. The public would have to pay a conservation tax either in the form of a small percentage increase on the general level of taxation, or as an increase on the 'check out purchase price'. Just as many regions of Europe currently add a 'visitor tax' then so the regions might apply a conservation tax to finance local conservation initiatives.

Traditional Attitudes

The attitudes of western civilisations have been shaped by our Judeo-Christian beliefs (McHarg, 1969). Nowhere is this more clearly seen than in our attitudes towards the way in which we respond to biosphere stimuli (White, 1967). We have been traditionally led to believe that the development ethic, in which mankind is master and the biosphere is a subservient provider of resources, is unquestionably the only relationship between man and our planet. The Christian religion, while teaching love and compassion to fellow men, is decidedly antagonistic towards all other components of our planet! (Barbour, 1973)

In contrast, many of the eastern religions, for example, the Hindu and Zen-Buddhist, take mankind merely as another animal which must live alongside other plants and animals. In some 'primitive' societies, religious and tribal kinship may extend beyond human members to include specific animals. Mossman (1974) has explained how some southern African tribal groups are organised on the basis of separate clans with specific animal totems. Thus, the members of the Dube (zebra) clan would not hunt its totem.

Western civilisations tend to be scornful of such attitudes, but totems and taboos have developed as deliberate relationships and have evolved over very long time periods to give advantages to both animals, plants and man. Inhabitants of the developed world may have much to learn from our fellow beings in the under developed world, not least, the means by which we too can learn the 'conservation ethic'.

Changing our Life-style

Let us assume for the moment that society has accepted the advantages to be gained from a conservation-based lifestyle; what changes would it involve, compared to our current patterns? Would it mean riding a bicycle instead of a car? Does it mean rejecting foodstuffs which have been produced by factory-farming systems in favour of organic foods? In both cases, the answer would be no. It might mean driving a different type of car, perhaps with an electric propulsion system, and a cruising speed of 90 k.p.h. instead of 120 k.p.h. It might also mean avoiding certain types of foodstuffs which are very expensive to produce in terms of land requirements and energy inputs, for example, the open range-land beef cow (Pimental et al., 1984). An expansion of recycling of commonly used domestic materials would also be appropriate. Besides saving essential raw materials it would allow the individual the chance to participate in the conservation policy.

The conservative-based life style would involve the careful selection of individual techniques for food production, waste disposal, transportation, heating of buildings, energy production and building design which both on their own, and in combination with one another, would make more efficient use of resources. Not only would manufacturing costs be significant. Running and repair costs and the ease with which components could be recycled would all be of major significance in the design of a conservation-based system.

Such a strategy would make great demands upon our technologists. Some opponents to conservation falsely claim that a life-style based upon conservation principles involves less technology and a return to the 'rural idyll' of our forefathers. Nothing could be less true. A conservation strategy with which to take our societies forward into the twenty-first century will demand the highest development of scientific and technological capabilities.

Ecosystems would have to be understood in a detail which has so far eluded us. Energy transfer between ecosystem components would be a key area of research as would the detail of population dynamics. Inter- and intra-relationships within and between plants and animals would require study. Complex cycling of materials and efficient use of energy would be of paramount concern. Gradually, we would work towards a built environment which approached the efficiency of the natural environment.

The achievement of a society run on sound conservation principles would represent a major conceptual and moral advance for mankind (Platt, 1966). At present, our societies are held in place by a basic animalistic response - that of fear (Westing, 1977). The militarist nuclear-strike deterrent as advocated by the superpowers represents at least a shocking waste of resources, at worst it could result in the annihilation of civilisation and our biosphere. A conservationist-based life-style represents a considerably better prospect for the long-term survival of mankind.

The Next Step

Before a conservation-based strategy for survival would be adopted by the developed nations a much clearer exposition must be made to national governments of the likely situation we could find ourselves in if we continue with our present exploitive systems.

Considerable advances have been made in the last twenty years. Much fundamental research has already been made, not only by ecologists and conservationists, but also in the mathematical modelling of ecosystems by applied physicists (Wilson, 1981). Engineers have been researching the design of natural systems, and have tried to copy some of these ideas. Media studies (film and television) have developed ways of informing the general public of the many ecological and conservation problems and solutions while at schools, colleges and universities, classes in conservation, ecology and environment are now integral parts of many syllabuses.

There still remains the challenge of convincing the politician that the conservation lobby is worth more than a few votes from the fringe interest groups in society. In Britain, a hesitant step was taken in 1972 to found The Ecology Party (The Ecologist, 1972). Its success was far exceeded by the emergence of the Green Movement in West Germany (Papadakis, 1984) and which now attracts about 10% of votes in that country. The image of The Green Party outside West Germany is still characterised by the archetypal long-haired, anti-nuclear 'hippie'. Within West Germany, the party is now accepted as representing the views of mainly young, professional people who are seriously concerned about the future ability of the biosphere to support mankind. Its influence is increasing both within West Germany and beyond and the values implicit in the Green Party may well prove to be the precursor of much greater changes in the ways in which western societies react to the biosphere.

The Problem Species - Man

One species has emerged to upset the biosphere. That species is *Homo sapiens*, mankind. Mumford (1966) has claimed that "our unique capability to combine a wide variety of animal propensities into an emergent cultural entity" has given us a distinctive *human* personality.

A consequence of *Homo sapiens*' success has been an acceleration in the rate of biosphere change. For most of our history those changes have been confined to a localised, short lasting effect. In more recent times, perhaps only as recently as the beginning of the twentieth century, our impact on the biosphere has increased due, in part, to our increased number, and also due to an increased ability and seeming desire, to create change for no reason other than that a dynamic changing human landscape is considered an indication of a vibrant society.

The effect of this last type of change on the biosphere has been to cause a rapidity and magnitude of change never before witnessed in the history of our planet. The constancy of change is also greater than before. The combined impact of these new changes has been to cause a *simplification* of the biosphere. Entire ecosystems have been destroyed along with countless thousands of plant and animal species (see Chapters 6 and 7). Feeding chains have been made shorter, and with fewer components, leading to a faster throughput of energy and materials. Some areas of the biosphere show clear signs of possessing less homeostasis than in the historic past. This is most clearly seen in the over-populated, drought prone areas of our planet. Many areas of extra-tropical Africa have suffered irreparable soil erosion as a result of deforestation to make way for agricultural systems (Owen, 1973).

Anthropogenic Change

It is easy to overlook the fact that our ancestors were insignificant by comparison with modern man in terms of their ability to 'manipulate' the ecosystems which surrounded them. Only gradually did man's capability as a strategic opportunist become evident. There are many other examples of plants and animals behaving as opportunists. *Epilobium agustifolium*, the fire weed, *Pteridium aquilinum*, bracken and Sturnus vulgaris, the starling, all exploit particular habitats and increase their numbers at the expense of others. Man is the strategic opportunist par excellence.

Ecological opportunists are often characterised by a behaviour pattern dominated by growth, greed and complete dominance over other species. The population curve of a strategic opportunist species usually shows a rapid and continual rise until the ability of the ecosystem to support such a large population size is exceeded and a population 'crash' occurs (Phillipson, 1971). Comparison of the population curve of the Kaibab deer in Arizona, Fig 1.1a with the world population for man, Fig 1.1b shows the typical pattern of population behaviour. Between 1906 and 1924 the Kaibab deer population increased by 96,000, following which a crash reduced the population in three stages almost to its 1906 figure (Rasmussen, 1941). The human population has not yet crashed despite its exponential increase but the parallel between the deer and human numbers is clear for all to see.

Fig 1.1a Kaibab Deer Population 1903-38

data from Rasmussen (1941)

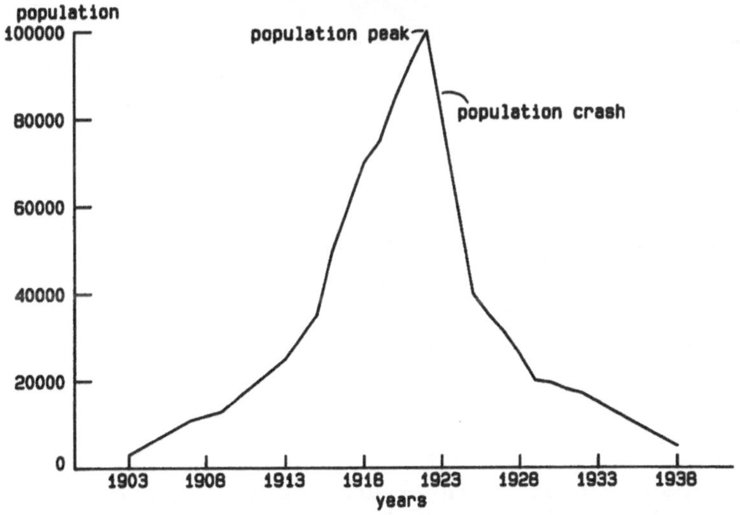

It is impossible to escape the fact that the most significant event to have taken place within the biosphere is the phenomenal growth of the human population. The increasing speed and the consistency of the population growth has resulted in the biosphere being placed under ever-increasing pressure to yield more of its resources to support just one species. More people mean more homes, more food, more jobs. Finite resources are used more quickly, the natural planetary cycling of materials becomes disturbed, pollutants accumulate in the biosphere, reducing the efficiency of its operation.

Fig 1.1b Population Data for Mankind 10000 BP to Present

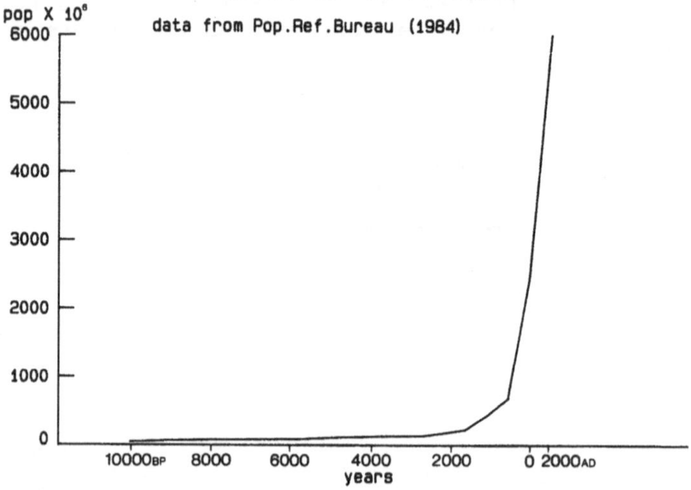

The Massachusetts Institute of Technology's Study of Critical Environmental Problems (MIT, 1970) identified that man's demand upon the biosphere in terms of resource demands and pollution inputs was doubling every 13.5 years (faster than the population doubling time at about 32 years). Not only are demand levels rising but the type of demand is undergoing marked change due to advances in scientific and technological discovery.

Malone (1976) suggested three new critical areas:

1. The hazards which are now apparent in the biosphere are increasingly caused by human activity.

2. Far from showing a greater understanding of what causes biosphere change, we are now less precise in relating cause to effect. This is due to improved understanding of the complexity and inter-relationships which exist in the biosphere.

3. Small scale, localised effects must not be viewed as one-off events. They can produce 'ripple-effect' on other ecosystems while some events can become cumulative.

The Role of Conservation

It is, perhaps, to be expected that in a book such as this that a strong case will be made for a conservationist attitude as a palliative for some of the crises that are currently evident in our biosphere. That conservation can 'cure' the malaise that now affects our biosphere is far too simple an assumption. Our species is too far along the technological track to turn back on what is commonly regarded as the 'technological fix' solution to biosphere problem solving (Commoner, 1969).

We must also recognise the spatial extent of many of the biospheric problems. As Malone (op cit) has stated a biospheric hazard in operation at location A may have an impact at point B. The distance between the 'cause'and the 'effect' may be separated by hundreds, or even thousands of kilometres, for example the transfer of radio-active nuclear fall out from Bikini Atol in the Pacific Ocean to New Zealand. How can a conservation policy in operation at point B have any effect upon the instigation of the problem at point A?

The enormity of the problem demands international cooperation at the highest government level. This often dissuades the individual person from making any personal effort to safeguard biospheric resources (Park, 1980). An analogy has been made by Hardin (1968) in which he described the haphazard growth of a primitive pastoralist community. Gradually, this society exploited its grazing lands until eventually, the system was broken and the society was doomed to destruction. The parallel is drawn with the treatment we make of our biosphere. The biosphere resource base is currently being made to increase its yield. Unlike the example quoted by Hardin (op cit) the biosphere has not yet collapsed although the significance of the example is clear. Each of the early pastoralist farmers thought only of their own direct well-being, little realising that the combined pressure exerted by all the farmers would one day destroy the system upon which they depended for their prosperity. Our use of resources shows a pattern in which constant growth is paramount and yet the supply of the resources is finite.

Hardin's scenario has been called the Tragedy of the Commons. It is one man's view of the way in which *Homo sapiens* has mis-used his resource base. Park (1980) has suggested a cautionary approach to the building of scenarios. It is too easy to construct a scenario on an emotional or a sentimental viewpoint and not upon a balanced scientific base (Clayton,1971).

Hodson (1972) has warned that the graphic descriptions, of which scenarios are usually made, may actually convey a situation which is so vast and complex that the individual citizen cannot contribute to its solution. This interpretation could be applied to Tragedy of the Commons. The situation in which the grassland system collapsed was brought about by the expansion and 'improvement' of the agricultural system. For every one collapsed system there were many others which survived. Exploitation of an ecosystem need not end in disaster - indeed, the present-day success of man as a species suggests that utilisation of ecosystems has been to our positive advantage.

The case for conservation is based upon those ecosystems which have failed the test of development. The conservation argument claims that all ecosystems, and eventually the biosphere itself, is liable to destruction if over-exploited. With a greater pressure now being placed on the biosphere the chances of collapse have increased. Most governments, irrespective of political motivation have accepted this argument to varying degrees. Acceptance on its own is insufficient. Research is needed so that optimal use of the biosphere can be achieved with a minimum of adverse side effects which are presently so common.

Conservation - A Symbiotic Relationship

Man obtains his basic sustenance from the environment which surrounds him. The passage of food, oxygen and water from the environment to man is accompanied by a return flow from man to the environment. The form of this flow is complex but is made up mainly of the waste products from man along with many abstract stimuli such as the effect of removing trees, ploughing soil, burning grassland. These processes make an impact not only on forests, the land and grasses but also on the countless other organisms which live in and on these systems.

No relationship takes place in isolation from the environment. It is possible to show the main ecosystem linkages by means of a process - response diagram in which arrows summarise the direction of stimuli movement. Such a diagram, Fig 1.2, is called an ecosystem diagram (Eyre,1970). In its simplest form it already contains eleven lines of relationship.

Fig 1.2 A Simple Ecosystem Diagram

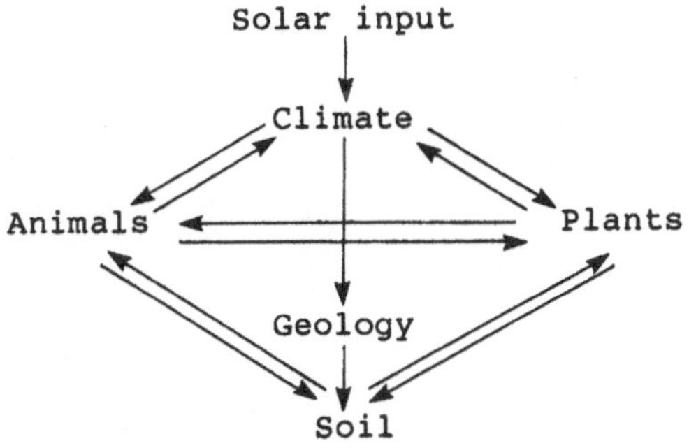

Here it is sufficient to recognise certain basic facts. These are:

1. *Homo sapiens* is an animal and as a consequence has finite biophysical requirements eg. temperature and gaseous requirements which must be satisfied if he is to survive.

2. *Homo sapiens* responds to environmental stimuli. These may be natural, e.g. climatic, famine, competition from other species. Man differs from other animals in that he also responds to complex cultural and social stimuli, e.g. love, fear, religion, political and economic conditions.

3. As a result of interaction between points 1 and 2 certain areas of the biosphere become better suited to support *Homo sapiens*. Richness and diversity of stimuli encourage the development of mankind, whereas poverty and uniformity of stimuli restrict development.

It is because the symbiotic relationship between man and the environment has fallen into disrepair or imbalance that the need for a conservation policy becomes a necessity. In order that conservation can be applied correctly we must first understand the ecological principles which govern the lives of plants, animals and man, and secondly we must study the distributional (geographical) patterns of ecosystems to be found on this planet.

Chapter Two

THE BIOSPHERE - OUR ULTIMATE RESOURCE BASE

DEFINITION OF THE BIOSPHERE

The biosphere is the habitat of all the life forms on this planet. Dasmann (1976) has defined it as:

> The fine layer of soil, rock, water and air that surrounds planet earth, along with the living organisms for which it provides support, and which modify it in directions that either enhance or lessen its life-supporting capacity.

Within the biosphere can be found all the myriads of different ecosystems. Each ecosystem interrelates with its neighbour, a change in one ecosystem will cause a ripple effect to run across adjacent systems, bringing an element of change.

For many hundreds of millions of years the biosphere underwent changes brought about by natural processes. Orogenesis, vulcanism and continental drift are examples of major causes of change, while the competitive forces between plant and animal populations brought about 'fine tuning' of separate ecosystems. Catastrophic environmental events, such as an ice age, brought equally severe changes in plant and animal population distributions and in population numbers.

At no time, however, was the continuation of the biosphere in jeopardy. Its survival was ensured by the complex network of interdependent ecosystems. Change was, and still is, an integral and characteristic feature of the biosphere.

It must not be assumed, however, that the biosphere has an infinite capacity to endure changed conditions. The latitude for change at any one point and at a given time is very small but through the persistent operation of changing environmental values over immense time scales, considerable change becomes possible (Cox et al, 1976).

Through the process of environmental change the biosphere has evolved into its present day shape, size and complexity. Moran et al (1980) have suggested that the contemporary complexity of the biosphere is greater than at any point in its long history. Complexity in the biosphere is considered an advantageous feature as complex inter-relationships between ecosystem components produce a strong, homeostatic (self-perpetuating) biosphere (Kellman, 1974).

Composition of the Biosphere

The biosphere comprises a unique combination of atmospheric gases, water, nutrients and solids which provide a habitat for all life forms on this planet. As far as physicists and astronomers can tell, our biosphere is unique to planet earth (Folsom, 1979). It certainly is not replicated in any of the other planets of our solar system although there may be similar combinations of gases, liquids and solids in other parts of our galaxy.

The biosphere is vital for a number of reasons:

a) It provides a 'safe' habitat for life forms in that it is the source of food and shelter for organisms. Within its protection all life forms fulfil their life history.

b) The biosphere has provided the functions outlined in (a) for approximately 1.4×10^9 years. The biosphere has not remained static but has fluctuated between extreme conditions. These extreme situations have not been inimical to life forms, indeed biosphere change may have assisted the evolutionary process.

c) The biosphere will remain the home of mankind and all plants and animals for the foreseeable future. It is possible that space travel will become a reality and will allow mass migration to a new and unused, biosphere. Until that time we must use our biosphere and its resources with care for without its protection life is not possible (Paludan, 1985).

d) The biosphere is a self-regenerating resource which draws its energy from the sun. Material resources are contained within the biosphere and are recycled on time scales ranging from a few days (water vapour and carbon) to many millions of years (sedimentary cycles).

The Biosphere - Its Component Parts

All plants and animals, including man, share the same ultimate living space. This living space is the 'biosphere'. The biosphere is an abstract space made up from parts of the three great earthly spheres, the lithosphere, the hydrosphere and atmosphere, Fig 2.1.

Fig 2.1 The Main Components of the Biosphere

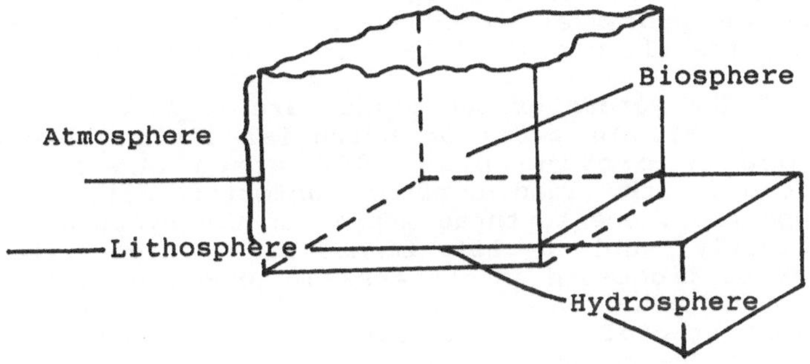

Lithosphere

Man spends the greatest part of his existence on the very surface of the lithosphere - the so called crust. The crust forms only 0.4% of the mass of our planet yet it assumes a significance out of all proportion to its mass. The crust is important in that it weathers to form the soil, a key physical resource necessary for life. Also, in the case of mankind, the lithosphere provides a wide range of 'resources' in the form of building materials, metalliferous and non-metalliferous ores and fossil fuels. The lithosphere also provides a surface upon and within which plants and animals can construct their homes, move about on (animals only), and fulfil their life cycle. Without a productive lithosphere it would have been impossible for our civilisations to have developed.

Hydrosphere

Resting upon the lithosphere and covering it to the extent of some 66% is the fluid hydrosphere. Most of the hydrosphere, some 97%, has a substantial amount of mineral salts dissolved within it. These minerals originate in the lithosphere. Only 10^{-4}% of the hydrosphere comprises fresh water (devoid of salts), 2% of the water is locked up as solid water (ice and snow) and 10^{-3}% of the hydrosphere is temporarily located in the atmosphere. The significance of water in our biosphere is of absolute importance; water molecules (H_2O) are an essential component of life on planet earth. The presence of water has allowed the formation of an atmosphere; it permits weathering processes to act upon the lithosphere and it supports all organic life forms (Goudie, 1984).

Most of the waters of our planet are located in the oceans (97%). In terms of potential living space, the oceans provide over 2000 times the space provided by the land surface. Unfortunately, only the uppermost one to three metres of the hydrosphere is directly usable by life forms. Below three metres a lack of light and an increase in pressure combine to provide unsuitable living conditions for all but the most specialised creatures. The great volume of the oceans represents a very useful depository for a whole range of man-made pollutants. The use of the hydrosphere as a dumping ground has many unresolved implications for biosphere conservation (Mellanby, 1967; McCaul, 1974).

Atmosphere

The final component of the biosphere is the gaseous atmosphere, Table 2.1. The atmosphere extends for about 32km but only 5.5km of this is directly usable by organisms. Beyond 32km the atmosphere gives way to 'space'. The atmosphere provides essential gases necessary for respiration by animals (oxygen) and plants (carbon dioxide). As for the hydrosphere, the atmosphere supplies very little by way of directly usable resources for man but its prime use is again a 'sink' area for air-borne pollutants, both gaseous and particulate in nature (Holdgate, 1979).

Table 2.1 Gaseous Composition of the Atmosphere data from Stoker and Seager, (1976)

Gaseous Components	Formula	Volume Percentage	Parts per Million
Nitrogen	N_2	78.08	780,800
Oxygen	O_2	20.95	209,500
Argon	Ar	0.934	9,340
Carbon dioxide	CO_2	0.0314	314
Neon	Ne	0.00182	18
Helium	He	0.000524	5
Methane	CH_4	0.0002	2
Krypton	Kr	0.000114	1

The Biosphere - Our 3-D Living Space

The biosphere comprises a three dimensional zone extending down into the hydrosphere about 3m, into the lithosphere for about 2m and extending upwards into the atmosphere for some 5500m. Within these orbits, conditions become uniquely suitable for the sustenance of organic life forms. Although it is possible to study the components of the biosphere in isolation it should be stressed that in reality those parts of the lithosphere, hydrosphere and atmosphere which form the biosphere become intimately interlinked. Thus the lithosphere has some water and air contained within it, the atmosphere contains water vapour and fine mineral particulate matter while the hydrosphere contains dissolved oxygen and carbon dioxide and suspended mineral matter.

Biosphere and Habitats

Within the biosphere can be found some distinct 'habitats' each comprising specific combinations of energy inputs, moisture availability and nutrients

(Willis, 1973). Each habitat can be likened to a three dimensional space. The shape comprises a two-dimensional area (length and breadth) which extends either/both upwards into the atmosphere and down into the lithosphere and/or hydrosphere. Finally, this three dimensional shape evolves and changes its volume with the passage of time, Fig 2.2.

Fig 2.2 Development of the Hypervolume Habitat Area

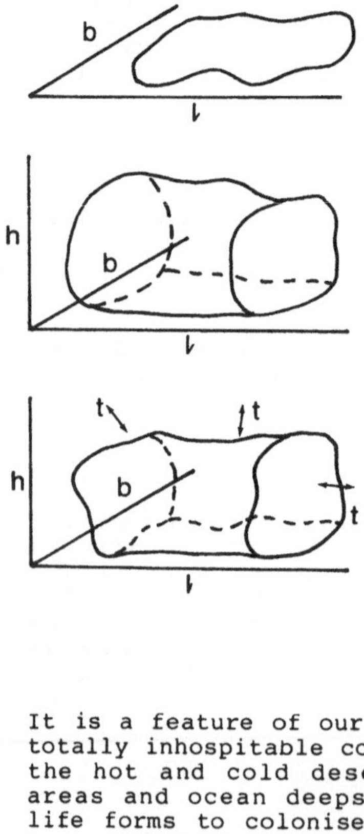

i. simple, two-dimensional area

ii. three-dimensional living space

iii. hypervolume area which changes its volume with the passage of time

It is a feature of our biosphere that there are few totally inhospitable conditions for life. Certainly, the hot and cold deserts, the ice-covered polar areas and ocean deeps are the most difficult for life forms to colonise but if a species can become hyper-specialised for survival in a very difficult region then its specialism can become an advantage and isolate it from competition from other species.

Having argued for the advantages of specialisation, however, it must be emphasised that the most successful species are the 'generalists' - species with tolerance of a wide range of different conditions.

Biosphere and Species Diversity

Those parts of the biosphere which are characterised by physical extremes are generally populated by sparse numbers of individuals and species. Conversely, where biosphere resources are abundant then species diversity and total numbers of individuals increase rapidly. This feature can be best illustrated by counting the number of species of beetles, or of reptiles that occur in distinct latitudinal lands, see Table 2.2. The varying capability of habitats to support a population is termed the 'habitat productivity' or 'species diversity' (Clarke, 1965).

Table 2.2 Increase in Species Diversity with Decreasing Latitude, North America. Data from Clarke, 1965

	Florida	Massachusetts	Labrador	Baffin Island
Beetles	4000	2000	169	90
Land snails	250	100	25	0
Molluscs	425	175	60	*
Reptiles	107	21	5	0
Amphibians	50	21	17	0
Freshwater fish	*	75	20	1
Coastal fish	650	225	75	*
Flowering plants	2500	1650	390	218
Ferns, mosses	*	70	31	11

* no data

Organisms have evolved so that they can complete their life cycles under the most extreme habitat conditions. The evolution may involve physical body change, for example the sole (Solea solea), a shallow in-shore salt water fish which has become flattened with a consequent movement of the left eye position until it comes to be close to the right eye (which remains in its original position). A restructuring of the jaws, and a change in fin and muscle positions compared with their free swimming cousins completes the physical changes of the sole. Physical change is usually accompanied by behavioural change, indeed it can be shown that it is persistent behavioural change which eventually leads to physical change (Parker et al, 1962).

Of all the species in the biosphere none is so adept at survival as mankind. What is easily overlooked is that whereas today, and for the past several hundreds of years, man deliberately strives to enhance his chances of survival, the original situation was very different. Our original hominoid ancestors showed no deliberate tendency to out perform the other plants and animals. Whatever happened all those millennia ago did so in an entirely voluntary manner.

The Uniqueness of Man

The fundamental change which occurred in a group of ancient primates was the emergence of a behavioural trait which persisted for a sufficient time such that a physical (skeletal) change occurred. The specific behavioural change was one in which instead of moving on all four limbs, proto-man gradually made less use of his fore-limbs for movement. A bi-pedal gait initially had many disadvantages compared to four-legged movement. It was slower, more clumsy and was a less well balanced form of movement. Its advantage was that it released the fore limbs for other functions (Bronowski, 1973).

Taken on its own, bi-pedal gait does not bestow any advantage - indeed the opposite. In order for the new form of locomotion to be useful a whole host of ancillary behavioural patterns become necessary and, of greater significance, some substantial neuro-physical changes must also take place. These have been summarised in Table 2.3. As with so many other lines of evolution, once a successful strategy has been devised then so it becomes difficult for that sequence to be altered.

What started as a chance behavioural pattern has today established mankind as the dominant organism on this planet. Such is our technological and scientific power it is difficult to imagine how, if ever, man will be replaced as the dominant organism.

Table 2.3 Changes Consequent Upon Bi-pedal Gait

Behavioural change - walking on hind legs
Neuro-physical changes necessary to optimise behavioural trait: 1. More neuro-muscular control over front limbs. 2. Greater manual dexterity of front limbs. 3. Improvements in sensory stimuli in hands and fingers (touch, feel, response to pain and heat) and better stereoscopic sight. 4. Faster neurological connections between hands, eyes and brain. 5. Enlarged cranial capacity to support improvements 1 - 4.

The behavioural and physical adaptations which have allowed us to achieve dominance over all other plants and animals have been accompanied by a single mindedness and ruthlessness in the ways in which we relate to other organisms. Many other animals (and some plants) show a similar determination to survive. These species are usually the ecosystem dominants, for example, the lion of the African savanna.

The main difference between the way in which ecosystem dominance has been achieved by man and that of other animals is that the latter achieve their dominance by working within the ecosystem boundaries. Man, because of his deliberate attempts to achieve dominance, has removed himself from the confines of the ecosystem. But man is still bound by animalistic requirements and is dependent upon the biosphere to provide the raw materials necessary for survival. It is the overlooking of this fact which has led man to over-exploit the biosphere and its components and which has given rise to resource depletion. It is for this very reason that conservation of the biosphere, its resources, ecosystems and species has become necessary.

It might be argued that <u>Homo sapiens</u> as a species has achieved dominant status specifically because of our non-compassionate attitudes to other species. A conservationist attitude has not been a traditional part of the human characteristic. If such an attitude were to be rigidly applied both now and in the future, then it might be sufficient to threaten man's position as the dominant species on this planet.

Survival Strategy and Conservation

It is because man is an intelligent being, capable of selecting an optimum strategy from a variety of opportunities that conservation of the biosphere has a chance of success. Conservation can be an emotive topic, involved as it is with ethics, behaviour, rights of non-human creatures, landscapes; but when stripped of such emotives, conservation becomes concerned with one issue - survival. Can man be so ignorant of his future well-being that he ignores the need to incorporate a conservation attitude into his life style?

A conservation policy based upon subjective, emotive reasons has no chance of success. On the other hand, conservation based upon well-founded, ecological arguments can present a logical and inevitable strategy with which to counter the trend towards the incessant growth which has so far been a hall-mark of man's civilisation. If a conservation attitude is to become an integral part of man's survival kit then it is essential that we understand what the biosphere comprises, how it influences our well-being and why some of the past actions of mankind have led to its disrepair.

Individual species, let alone individual organisms, make use of only very small and usually specific parts of the biosphere. The amalgamation of the individual components of atmosphere, lithosphere and hydrosphere provide distinctive physical conditions which comprise 'habitats' which in turn, can be colonised by specific groupings of plants and animals. All the species which exist in habitat 'A' will have similar requirements for survival. Some of the inhabitants of habitat 'A' may also be able to survive in different habitats e.g. 'B','C' or 'D' but it is extremely unlikely that <u>all</u> inhabitants of

'A' will be able to find <u>one</u> other mutually acceptable habitat. In other words, if habitat 'A' is destroyed then the particular grouping of organisms in 'A' will be lost. Individual species may also be lost if no suitable alternative habitats are available.

To summarise: biosphere change (whether natural or induced by man) can bring about several different results:

1. If a specific habitat is destroyed the species agglomeration within that habitat is lost.

2. Species from the lost habitat can regroup (but in different combinations) in different habitats.

3 If no suitable local habitat exists then species may become 'homeless' and they cease to exist (they become extinct).

ECOSYSTEMS - THE BASIC BIOSPHERE MODEL

The different habitats which exist in the biosphere support various combinations of life forms. The ways in which life forms interact with one another and with their habitat surroundings can best be studied by use of a model called an 'ecosystem'. Because of the almost infinite variety in the habitat conditions then so there is an equally wide variety of ecosystem types.

The term 'ecosystem' has been used for about 70 years, the first written reference to the word being in Clements (1916) though it was Gleason (1922) who made extensive use of the term to infer the unique combinations of plants, animals, climate and soils. Tansley (1935) also made extensive reference to the term in his study of relationships between vegetation and habitat conditions.

Eyre (1970) was responsible for renewing the bioecologist's interest in the ecosystem. It was he who portrayed the ecosystem as a concept or model which allowed the organism and its environment to be studied as an integrated unit, Fig 1.2. Because ecosystems are merely models they cannot exist in reality. In this respect they differ from habitats which do exist in reality. The habitat comprises the physical and chemical components which support life forms.

Types of Ecosystem Models

Ecosystems exist as conceptual models in the minds of ecologists. They can be transferred into written form (as a descriptive model), or more commonly nowadays as a series of mathematical algorithms for use in a computer program. The value of the ecosystem model can be found in the way in which it can mimic the relationships between the living and non-living components of a habitat.

Whereas the ecosystem is a model, the ecosystem components (plants, animals, soil and climate) exist in real life. They can be measured, captured and studied in detail. When treated in this way a detailed ecosystem description can be made but there would be little by way of understanding the means by which the components were interrelated.

The ecosystem concept is an attempt to allow the ecologist a method of studying and hence understanding the relationship between the living and non-living components of a habitat. This is a difficult task and is possible only for relatively simple habitats. Chorley & Kennedy (1971) have suggested that ecosystem study can be made at three different levels:

a. 'Black-box' level in which system inputs and outputs can be identified and measured but the inter-relationships which take place between components are known; this is probably the most common situation in habitat studies.

b. 'Grey-box' level: some of the connections between components are understood though substantial areas are still poorly understood. This level of understanding can best be seen in habitats which have been constructed (or repaired) by man, eg. agricultural systems, reforestation, reclaimed land, sand dune stabilisation.

c. 'White-box' level of understanding implies a complete knowledge of the habitat relationships. Habitat behaviour can be predicted with accuracy.

This situation is still uncommon and is confined to the simplest of habitat situations. The work of Teal (1957) and Duvigneaud (1971) illustrates the complexity of this level of understanding. Computer models and simulation studies (May, 1974; Horn, 1981), have enabled progressive advances to be made in recent years.

Usefulness of the Ecosystem Concept

There is little doubt that the ecosystem model can be of great use in improving our understanding of individual habitats and the biosphere (Jones, 1983). However, applying the ecosystem concept to real-life situations is difficult. If the ecosystem concept is applied at the 'black-box' level then little by way of real understanding of how the habitat functions can be achieved. On the other hand, to work at 'white-box' level demands a small team of research workers, sophisticated computer capabilities (massive computer storage and an ability to write the necessary programs), a time span of perhaps a decade and finance to support the research effort.

It is appropriate to apply the concept of 'successive approximation' as developed by Poore (1964) to the study of ecosystems. Poore, op cit, suggested that when working with vegetation community analysis it would be necessary to begin at the simplest level and to gradually increase the intensity of study so that greater understanding of the community could be obtained. A similar process of 'successive approximation' is directly applicable to ecosystem studies. Initial work would be made at the black box level. Gradually, and with increasing experience, a move towards grey-box level could be made. Eventually, and for a small selection of ecosystems, it should be possible to work at white box level.

Ecosystems and Environment

The ecosystem is a complex interrelationship of abiotic components (the climate, geology and soils) and biotic components (the plants and the animals). These components do not exist in an isolated state but react to stimuli which may be produced from within the ecosystem, or from other adjacent ecosystem or from the surrounding environment.

The term 'environment' has been used by Clarke

(1965) to imply the space which surrounds an individual organism, or groups of organisms. It is the hyper-space in which the total ecosystem plays out its existence. Just as the ecosystem was shown to be a conceptual model, p. 33, then so too the environment is an abstraction which has been devised by ecologists to assist the study of ecosystems.

Usually, a large number of different components combine to form the environment and, as a result, a reaction process occurs between the environmental inputs and the organism. A series of response signals are sent back from the organism to the environment and these signals can then modify the environment. According to Mason and Langenheim (1957) an environment can only exist when centred upon an organism and is a "reaction system of the organism and the phenomena which directly impinge upon it to affect its mode of life at any time throughout its life cycle". From this definition we can assume that phenomena which do not impinge upon an organism may be classified as 'non-environment' while others, which may at some future time be involved in the relationship, may be termed 'potential environment' (Kellman, 1974).

Although the environment may be abstract, its components are real. The individual parts of an environment do not exist in isolation but are interrelated in many different ways. In both these respects the environment and ecosystems are very similar; both are abstract terms but comprise real components, both comprise many different parts all of which are interrelated and inter-dependent for their well-being.

It is convenient to divide the total environment into its main physical, chemical and biological components, Table 2.4. Having studied the environmental components as separate parts it is then possible to reassemble the environment and to examine its 'wholeness'.

The Medium

Two fundamentally different media exist in the biosphere; these are the media of air and water. All organisms must possess a medium and, once selected, it is unlikely to be changed (Clarke, 1965). It is difficult to accept that in spite of all the different habitats to be found in our biosphere,

Table 2.4 The Major Components of the Biosphere

Physical	Chemical	Biological
The Medium	Properties of the Medium	Relationships between Species Types
The Substratum	Chemistry of the Substratum	Relationships within Species Types
The Climate		

from arctic wastes to tropical coral reefs, from arid, hot deserts to cold, acid oligotrophic swamps and from fresh water mountain streams to warm, polluted effluents, the medium in every instance must be either air or water.

This limited choice between media conveniently divides our habitats into two forms:

a) Terrestrial habitats in which the medium is air.

b) Aquatic habitats in which the medium is water; this habitat can be further subdivided into fresh water and salt water habitats.

The medium is the most basic requirement for all organisms. Plants and animals must possess a respiratory system which can function in one or other medium. Those organisms which use a medium of water, or a water-based fluid, rely on gills or trans-tissue diffusion in order to obtain oxygen (for animals) or carbon dioxide (for plants). Terrestrial life forms use a medium of air and have a more complex pulmonary respiration system.

Apart from the vital respiratory role, the media are also responsible for the provision of buoyancy or support. In this capacity, water is approximately 850 times more buoyant than air, Table 2.5. Accordingly, organisms which are confined to a medium of water can dispense with a rigid support mechanism. The 'bones' of a fish serve merely as a means of muscle attachment and provide little or no structural support. By comparison, a terrestrial animal requires a massively calcified bone structure to provide rigidity and onto which muscles can be attached. Land plants attain a vertical posture by means of extra cellulose or lignin in their cell walls whereas water plants dispense with such features, and have instead air spaces which serve as buoyancy tanks.

Table 2.5 Supportive Capacity of Air and Water Media

Medium	Density Value (g/cc @ 4°C and mean sea level)	Medium:Protoplasm Ratio
Air	13.0×10^{-4}	1:0.001265
Sea water	1028×10^{-3}	1:1
Fresh water	1001×10^{-3}	1:0.9737
Protoplasm	1028×10^{-3}	

The medium also has a major influence on the ease of movement. This is particularly relevant for animals and also assists the dispersal of plant fruits and seeds, Fig 2.3. A medium of air, because of its lower density, allows faster movement for less expenditure of energy than does a medium of water.

The Substratum

The substratum provides a surface or substance upon or within which an organism lives. It is sometimes difficult to distinguish where the medium ends and substratum begins. For example, the earthworm spends its entire life history in the soil; the substratum is soil while the medium is water. Surrounding the body of the earthworm is a watery fluid across which gaseous diffusion takes place. If the earthworm were artificially dried, then death would quickly result. Without the correct medium most organisms die, but if an earthworm was placed in a box containing damp polystyrene granules (a new substratum) then the worm could still live provided there was also a

Fig 2.3 Dispersal Adaptations to Suit Different Media

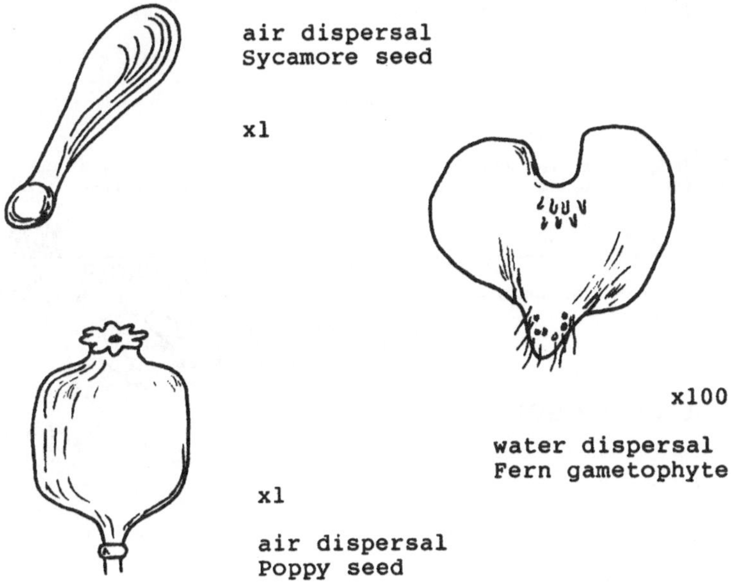

supply of nutrients. This illustrates a commonly observed feature; an organism can rarely change its medium while a change of substratum is quite common.

The most common source of substrata are rocks and rock derivatives, that is soil. Other substrata include wood, hide, leaves, roots and water. Most substrata can be used by different organisms to provide a variety of living areas. Fig 2.4 shows some of the ways in which water can serve as a substratum.

The function of the substratum has been defined by Andrewartha <u>et al</u> (1954) as providing:

1. A place of attachment (ie. a home).

2. A source from which nourishment (food) can be obtained.

3. Provision of shelter.

Fig 2.4 Water as a Substratum

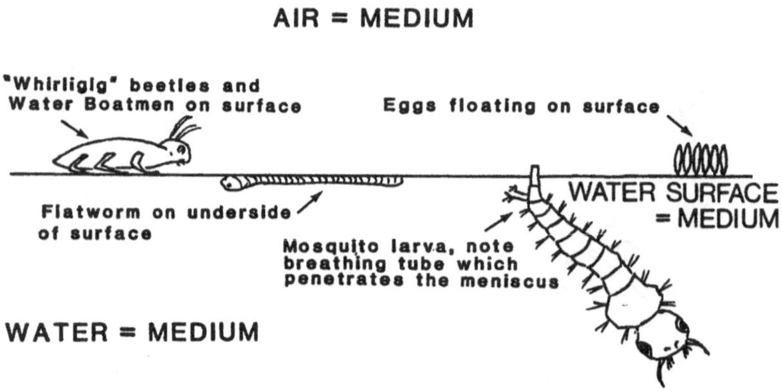

The Climate

Two climatic inputs, those of solar energy and precipitation, are fundamental requirements for life. From an ecological standpoint, climatic inputs can be considered an amalgam of favourable and unfavourable stimuli which promote or retard growth respectively.

The ecologist need not be concerned with the major atmospheric processes which determine the climate of an area. Instead, we must establish which critical climatological thresholds have relevance on the biosphere. In the past, climatologists have suggested that specific climatological values can be used as 'deterministic switches', the attainment of which causes a particular organic function to occur (Miller, 1961; Thornthwaite, 1954). Well established critical values such as $0^{\circ}C$, $6^{\circ}C$, 500mm precipitation per annum and 200 frost-free days per annum have all been variously proposed as critical threshold values for plant growth. More recent workers (Krebs, 1972) have suggested that critical values should not be considered as absolute values at which specific events occur. Instead, they should be used as marker points at which changes in ecosystem functions take place. Evans (1963) has provided examples to show how the quality, intensity and duration of climatic inputs can control the life history of plant species while Vernberg et al, (1970) give similar evidence for animal species.

Chemical Properties of the Medium

The proportions of two atmospheric gases, oxygen and carbon dioxide, are of critical importance for the support of life. Neither are thought to be in limiting supply even though carbon dioxide forms only 0.03% of the atmosphere by volume percentage. Stoker et al (1976) have explained that the general availability of CO_2, despite its small percentage occurrence, is due to the high mobility of carbon in the planetary cycling of this component. Some fears have been expressed by Manabe and Wetherall (1975) that the CO_2 content of the atmosphere could increase due to the burning of fossil fuels by man and the consequent release of CO_2 to the atmosphere. This, in turn, could result in a warming of the lower atmosphere by 0.5^{o}C by the year 2000 AD.

The chemistry of the media has been relatively constant probably for as long as 3.4×10^8 years (since Silurian times). From the early 1950s mankind's increasing use of high temperature and high pressure combustion processes along with the use of increasingly complex chemical substances has led to a massive increase of air and water pollutants. Sulphur dioxide, hydrogen sulphide and the nitrous oxide group of pollutants have become problem substances in the air over many of our industrialised conurbations. Hickey (1971) has stated that conclusive proof about what is unhealthy about pollutants is often unclear. The controversy over the source, the effect and the hazardous nature of acid precipitation (Hutchinson et al, 1980) is an example of the confusion which exists about the pollutants man has placed in the atmosphere.

The water contained within the biosphere is also

becoming contaminated by toxic chemical materials (McCaul, *et al*, 1974). Natural, unpolluted precipitation should have a slightly acid pH of about 6.5 (pH 7.0 = neutral). Rain-water contaminated by oceanic spray or precipitation which occurs during lightning storms is usually more acidic (pH 6.0), Edmonson (1971).

Mildly acid precipitation has some effect upon soil fertility in that the free hydrogen ions in the rainwater leach the chemical bases from the top layers of the soil. Only when this leaching process has continued for several thousands of years does evidence of soil nutrient exhaustion become evident (Cruickshank, 1972).

The impact of man's increasing trend towards industrialisation has led to an intensification of atmospheric acidification so much so that rainwater pH over urban/industrialised areas may now attain pH 5.5 (Goudie, 1984). The variety of dissolved acids has also increased. As a consequence, very acid precipitation can cause 'burn' marks on vegetation surfaces, can damage the vigour of plants and in extreme cases can cause the death of vegetation for example, Western Hemlock conifers (*Tsuga heterophylla*) planted on the Pennines in England (Hepting, 1964; Roberts, *et al*, 1983).

The chemistry of surface water bodies is further changed through the additional dumping of industrial and domestic wastes from which solutes migrate into rivers and lakes. The temperature of river waters can be raised by the addition of cooling waters from thermal power stations (Hodges, 1977).

Toxic Chemicals and Biosphere Catastrophes

We have so far averted a major change in the chemical composition of the atmosphere, not by any foresight on behalf of man, but by the ability of the atmosphere and hydrosphere to self-cleanse themselves of toxic materials. This is achieved mainly by the major cycling of materials between 'available' and 'non-available' reservoirs (Strahler *et al*, 1974).

A number of localised catastrophes have occurred, for example the Meuse Valley air pollution incident of 1930 and the smog which killed an estimated additional 4000 people in London in 1952 (Stern,

1968). The Los Angeles smog problem and similar problems in Tokyo and Mexico City have produced well documented effects on human health (Report on Air Pollution, 1970; Stewart, 1979). The Council on Environmental Quality (1973) have calculated the cost of pollution damage in the USA. By 1968 this figure was calculated at $16.2 billion of which $5.0 was credited to material and vegetation damage.

Chemical Composition of the Substratum

The chemistry of the substratum rarely becomes a major limiting factor on the way in which the biosphere functions. This is due to three basic reasons:

1. Organisms show a wide tolerance of substratum chemistry. Only the very young and the very old members of a population exhibit very specific responses to chemical excess or deficiency.

2. Apart from some exceptional geological outcrops (notably those containing high levels of the so-called 'heavy' elements such as lead, zinc and copper), the chemical composition of the surface layers of the lithosphere do not contain toxic levels of chemical elements.

3. By a process of trial and error, any organism which comes in contact with a reactive chemical surface may:

   a) migrate to a more favourable area,

   b) adapt to the chemical environment,

   c) if neither a) nor b) is possible then the population of that organism will be eliminated.

The influence of the chemistry of the substratum is shown in a number of different ways. Plants exhibit a particularly obvious response to soil chemistry conditions. Most plants require the presence of a large number of trace elements in the soil. There are 16 of these, Table 2.6. Not all are required throughout the life history of the plant.

Some plants show the ability to tolerate the absence of some trace elements provided that others are in unlimited supply. Of all the essential nutrients, nitrogen is probably the one which is most frequently in limiting supply. Nitrogen is a highly mobile element and is rapidly leached from the soil after periods of heavy rain. Nitrogen levels

Table 2.6 Trace Elements Necessary for Plant Growth from Buckman & Brady, (1970)

Macro-elements		Micro-elements	
Carbon Hydrogen Oxygen	atmosphere	Boron Copper Iron Manganese Zinc Molybdenum Chlorine	soil
Nitrogen	from fixation		
Phosphorus			
Potassium			
Calcium			
Magnesium	from soil		
Sulphur	from soil and atmospheric pollution		

can be reinstated by the activity of the resident soil nitrifying bacteria.

Pratt (1965) has shown that many of the other essential plant nutrients also become limiting due to soil wetness or dryness, extreme pH values or oxygen deficiency. Agricultural management techniques have now been developed to overcome most of these temporary deficiencies. The commonest method of alleviation is the application of inorganic fertilisers but this is a short-term and expensive process. A fuller understanding of soil chemistry has already shown that judicious timing of cultivation can aid the release of inherent soil fertility.

Relationships Between Species Types (Interspecific Relations)

Organisms come into contact for a whole host of reasons. On occasions the contact can be deliberate as when one species preys upon another. Elsewhere, contact may be fleeting and accidental as when two fish in an aquarium can be seen to almost collide, only to swerve and swim apart. Contact can be aggressive, cooperative or submissive. Contact can be of benefit to both parties or to one alone, Table 2.7.

Few organisms are totally solitary in their behaviour. Even the most reclusive animal such as the south-east Asian tree shrew of the Tupaia genus, a solitary, nocturnal forest dweller has to come in contact with plant species in order to gain a food supply. Feeding relationships are responsible for the majority of contacts between organisms. Decomposition and cleansing processs bring other relationships into being. Direct exploitation of a species by another is a rarity, the best known example must be the behaviour of the adult female cuckoo which deposits one egg into a host nest. Another example is the so-called 'slave-master' ants of the Polyergus genus who capture members of the Formica group of ants and put them to work building nests and collecting food. For the most part, exploitation is a temporary phase of behaviour and is shown only by organisms which are located at the apex of long and complex food chains.

These species, mostly aggressive carnivores such as the pike, golden eagle and lion, cannot be guaranteed a consistent supply of food and have become adapted to hunt down its prey with little signs of mercy.

Relationships Within Species Types (Intraspecific Relations).

This type of relationship has developed for three specific reasons:

1. To facilitate feeding - determined mainly by the minimum food supply available in the most unfavourable part of the year (in the dry season of winter period).

2. To aid reproduction - necessary for the bringing together of the sexes and hence the propagation of species.

3. For defensive purposes especially during times of adversity (at birth, during food shortage or during invasion of territory by

other species).

Intra-specific relations have evolved over countless generations and are often accompanied, particularly amongst the higher animals, by a complex social hierarchy and complex behavioural patterns. Within species relations are usually based on groupings of

Table 2.7 Association Types Possible Between Species
data from Burkholder (1952)

Species A	Species B	Relationship Type	
+	+	Mutualism	symbiosis
+	O	Commensalism	
O	O	Tolerance or neutrality	
O	-	Antibiosis	
+	-	Exploitation/parasitism	
-	-	Competition	

+ = a beneficial relationship
- = a declining or harmful relationship
O = neutrality, neither loss nor gain

individuals. Man behaves in true animalistic form in that we are highly gregarious beings who prefer to group together in large numbers (and by so doing have created villages, towns, cities, conurbations).

Plants too can suffer from overcrowding. Excessive seedling germination can produce too many plants per finite growing space and this excess becomes increasingly problematical as the plants grow older. Substantial species mortality of individuals occurs as a result of this competition.

THE CONCEPT OF WHOLENESS

This chapter has, until now, been concerned only with the identification of ecosystem components. A summary of these components has been given in Table 2.4. The remainder of this chapter will examine the ways in which these components fit together and function as a whole.

When the environment-ecosystem components become inter-connected by means of energy flow, material cycles, feeding patterns and inter-connected territories, the system assumes an additional, and unquantifiable dimension. The dictum 'the whole is more than the sum of the individual parts' can be applied to the undisturbed environment - ecosystem relationship. This is the concept of 'wholeness' and was first recognised by Aristotle (384-322 B.C.). It is of particular significance when studying environment-ecosystem relationships.

The Aristotelian dictum has been used by Von Bertalnffy (1962) in his development of 'system theory' as a mechanism for problem solving. A system theory approach to problem solving is one in which a concept or problem is examined in a specific manner. The problem is solved only through the asking and solving of relevant questions and the feeding back of the solution to the original question. The provision of a solution may have altered the initial hypothesis and if so, the process of question, solution and feedback must be repeated until an answer is obtained which can satisfy all cases of the problem.

The systems approach to learning is of particular relevance to the understanding of biosphere problems. Unfortunately, the systems learning approach is taught in very few schools. Instead, we are taught to ask the question 'why?'. This is Galileo's 'resolute' method of learning in which knowledge is broken down into smaller and smaller units. We become experts in small unit subjects (the systematics) but fail to understand the relationships between the units.

Biosphere Systems

Three main types of system can be found on our planet:

a. isolated systems

b. closed systems

c. open systems

It is with the last of these with which we are concerned. Chorley et al (1971) have listed the properties of an open system as including the free exchange of energy and matter across 'open' boundaries, the approximate balance between the input and output of both energy and material which in turn produces a system which is in a state of balance (homeostasis) and a system orderliness characterised by low entropy (highly organised).

Open systems can be further classified depending upon their internal character. Table 2.8 shows the divisions of such a classification along with some ecosystem examples. Deterministic open systems are now known not to occur in the biosphere, although in the early years of the present century the school of geographical determinism, as advocated by the German geographer Ratzel, argued strongly in its support (Wooldridge and East, 1962).

We know that biospheric systems work in an unpredictable way, they are 'probabilistic' systems in that no end result can be guaranteed. Fortunately, there are ways in which events taking place within probabilistic systems can be predicted; these methods involve the laws of probability. For example, if two adjacent agricultural grassland fields were allowed to return to natural vegetation it is exceedingly probable that after a given time period both fields would show a broadly similar appearance though in detail the species composition and structure would differ. The operation of probabilistic control mechanisms would ensure that, in their detail, the two fields would have reached different end points.

Ecosystem Control Mechanisms

If probability theory alone was responsible for directing the fortunes of the ecosystems to be found in our biosphere then ecological pattern and the forecasting of ecological events would be an impossible task. Instead, ecosytems are 'guided' by external and internal control mechanisms, more appropriately called the science of cybernetics

(Beer, 1967). Cybernetic control is widely used in industrial production lines. In its simplest state a cybernetic control is an on/off switch, for example a thermostat or a set of traffic lights which regulates the flow of traffic at an intersection. The cybernetic system can be refined, so that the traffic light example is extended to include a timing cycle, 30 seconds 'go' for main route A, 20 seconds for secondary route B and 10 seconds for pedestrian crossing. Or the system may have a traffic counter sensor or a photo-electric cell to detect the presence of traffic to help optimise traffic regulation. Even when taken to its ultimate development in which a central computer takes control of the entire traffic regulation via traffic lights, the system will remain a deterministic system because the programme of events will be controlled by the computer program itself. Only gradually are we learning how to program artificial intelligence (AI) into industrial robots but even in these examples the 'intelligence' is pre-programmed and is not 'learnt' in the organic sense.

Table 2.8 Structural Classification of Open Systems

Complexity level	Deterministic	Probabilistic
Simple organisations	Not represented in open systems which involve organisms	Simplest ecosystems e.g. inter-tidal rock pools, agriculture
Complex, but describable		Most extra-tropical ecosystems, e.g. taiga, deserts, deciduous woodlands
Complex, indescribable		Tropical ecosystems e.g tropical rain forest and the biosphere as a whole

Because natural variability is an inherent property of living creatures the organic system will never be deterministic in its behaviour. Lack (1946) provides a good example in his well known book about the behaviour of the common European robin (Erithacus rubecula). The number of breeding pairs of robins fluctuated in a given area from year to year. This was due to a number of environmental changes.

1. The robin population will increase under conditions of favourable (mild) winters and abundant food supply.

2. A succession of favourable breeding seasons produces an exponential increase in numbers of robins, and conversely, unfavourable conditions maintained for several years leads to a rapid depletion of numbers.

3. As the robin population increases then so individual territory size decreases leading to a smaller area from which food can be obtained. This in turn can lead to under nourishment and a reduction in the number of chicks reared per breeding cycle.

Fluctuations in population size of robins tends not to oscillate wildly between very large and very small numbers. Instead, the population cycles around a mean value which may show a 10% annual variation. Considering all the possible causes which could create either a gain or a decrease in population size, the inter-year variation is surprisingly small.

Constancy in Ecosystems

The example of the relative stability of population numbers as displayed by the robin illustrates a commonly observed feature of many ecosystems and species, that of constancy. Constancy can be measured in a number of different ways, for example, by counting the number of individual organisms in an area, or by calculating the biomass, or by noting the dates at which significant stages of the growth history are reached. All these indices tend to confirm the feature of constancy, that is, individuals show only a small variation around the mean value.

Changes will take place, see pages 23-24, but will

occur within boundaries imposed by feedback signals which the ecosystem and species will provide themselves.

The Feedback Mechanism

The success of an ecosystem is dependent upon the development of an accurate and speedy feedback mechanism. It can be argued that an agressive, expanding excosystem will replace an adjacent and less successful ecosystem because the former has a better developed feedback mechanism than the latter. Any organism which responds quickly to a stimulus is better placed to survive than an organism with lethargic reactions. An ecosystem which has developed a network of accurate and fast feedback mechanisms will display several characteristics, all of which equip that ecosystem to outperform the system with poorer feedback responses. Good feedback channels permit the ecosystem to maintain input levels as near as possible to optimum levels. To achieve this, the ecosystem will be capable of fine-tuning the inputs so that the amplitude of change rarely creates instability within the system.

Absolute stability is rarely found in an ecosystem although it is possible for an ecosystem to attain relative stability. In this condition, oscillation will occur between two extremes.

The system inputs can be considered to fluctuate as a series of independent wave-like curves and only when the majority of inputs come 'in phase' can stability be attained.

The condition of relative ecosystem stability is not commonly found in the biosphere at the present time. This is due to the process of simplification brought about by man, see Fig 2.5. Removal of species by hunting or by habitat destruction has resulted in most ecosytems now existing in a state of permanent instability. The ecosystem balance which has been achieved over tens, or even hundreds of thousands of years has been swept aside in just 10,000 years of time during which mankind has achieved biosphere dominance.

It is because of the magnitude and speed with which ecosystem simplification is now taking place that it has become necessary to actively promote the need for ecosystem and biosphere conservation. Had it been possible for mankind to be less destructive in his use of ecosystems then much of the present concern with conservation would have been unnecessary.

Fig 2.5 Example of Deciduous Forest Ecosystem Simplification

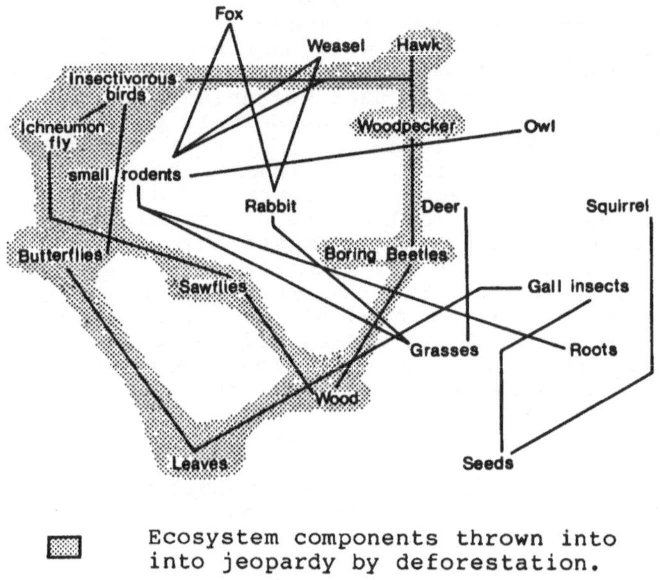

Chapter Three

MAN AND THE BIOSPHERE

Man's Initial Development

Homo sapiens is a relative newcomer to planet earth. Set against the entire time span over which the planet has existed our species has occupied only 0.01% of that time, see Table 3.1

Table 3.1 Critical Events in the History of Planet Earth.

Event	Date Before Present
Start of Solar System	4.6×10^9
First Traces of Free Water	4.4×10^9
Self-Replicating Molecules Appear	4.1×10^9
First Signs of Fossil Cells	3.6×10^9
Free Oxygen Appears in the Atmosphere	2.0×10^9
Complex Cells with Distinct Nuclei	1.5×10^9
Formation of Atmospheric Ozone Layer	1.4×10^9
Appearance of 'Modern' Flora and Fauna	45×10^6
Appearance of Proto-Man	45×10^4
Development of Agriculture	9×10^3

It is impossible to provide absolute answers as to when and where mankind first appeared. Nor can we ascertain with complete certainty the origin of our ancestors. All that can be said is that man's early history appears to have begun sooner than most anthropologists originally thought possible. Current evidence suggests our ancestors first appeared in the continent we now call Africa (Pfeiffer, 1972). Several sites, see Fig 3.1, for example the famous Olduvai Gorge (Leaky, et al, 1964), in Tanzania along with sites at Taung, Sterkfontein and Swartkrans in South Africa (Pilbeam, 1972) have revealed the remains of Australopithecus, an ape-like creature which walked erect, had a small brain box but whose jaw movement was very non-hominoid in that the chewing action was rotary and not up and down. Australopithecus probably had little ecological impact upon his environment. He was a herbivore with little or no technical skill.

Fig 3.1 Main Locations of Australopithecus in South and East Africa
from Tobias (1967)

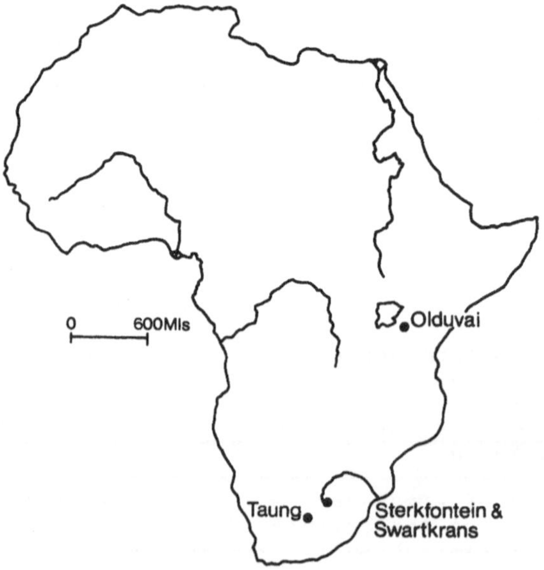

More recent interpretations from archaeological research at Swartkrans in the northern Transvaal of South Africa (Brain, 1967) has suggested that by about 450,000 years ago Australopithecus had disappeared and that two distinct forms of hominoid had emerged, both living alongside one another. One form had a massive jaw structure, large molars suitable for grinding food, large attachment points for muscles to work the jaws and a flattened cranium. It has been assumed on the basis of skull characteristics that this form of early man was a direct descendant from Australopithecus. It probably fed entirely on seeds, leaves and grasses.

The companion hominoid differed considerably in its bone structure and by implication, its life style too was different from the previous form. The lower jaw structure was narrower and much less calcified. The teeth were smaller and sharper and muscle attachment points were less developed. The overall shape of the skull differed in that it was 'high domed' suggesting an enlarged cranium. Anthropologists at the Transvaal Museum in Johannesburg have reconstructed the life style of this creature. Feeding habits were probably omnivorous (it was a scavenger); the presence of bone digging tools alongside skeletal remains imply that roots and tubers were dug out of the hard ground. Food was actively searched out and not merely collected in an involuntary manner. This species has been described by anthropologists as living 'by its wits'.

Revised interpretations of early man's development (Leaky, 1982) has placed greater emphasis on the use of brain superiority rather than the earlier theories which gave emphasis to the destructive and aggressive nature of man's character (Mumford, 1962; Nicholson, 1971, 1973). There would undoubtedly have been frequent physical contest with other animals and perhaps also with other human groups, though the small population size would have ensured relative isolation of family groups.

First Impact on the Environment

Australopithecus would have had minimal impact upon his surroundings. He was non-technological and numerically insignificant. Even some 450,000 years ago early man still had no clothing, no fire, and only the simplest of digging sticks while the only shelter available was that provided by natural caves or from overhanging vegetation. Pollution might have been a local problem, created mainly from human excrement and domestic wastes but early man was controlled by the main seasonal climatic changes and probably moved in harmony with the rain belts which migrated annually between the tropics. His migratory life style enabled him to leave behind the pollution which surrounded his hearth.

Constant movement placed early man in an environment of ceaseless change. Adaptability to new circumstances was a hallmark of our ancestors. The searching out of food, the navigation of routeways, the general art of survival all provided Homo sapiens with one advantage over all other species - it stimulated the development of an enlarged cranial capacity. No longer was instinctive behaviour the sole means of survival. Learning, cunning and mobility were three attributes which allowed man to develop a unique life style.

The hunting-gathering life style has dominated man's time on this planet. Miller and Armstrong (1982) have claimed 90% of all human beings that have ever lived have been hunter-gatherers, with 6% being settled agriculturalists and just 4% involved in industrial societies. It is primarily the occurrence of the latter group's activities within the last century which has created the need for conservation.

A hunting-gathering life style usually has little damaging impact on the environment. This is due to the low density of population which can be supported by this life style. Schwanitz (1967) has estimated that a hunting-gathering community requires about $20Km^2$ of land per person in order to guarantee year-round survival. A similar are of fertile land could, under a moderate agricultural system, support 60,000 people (Jones, 1979).

The circumstances in which early man existed would have shown ceaseless change and the mere act of survival would have been a major challenge for mankind. That the challenge was met and responded to in a highly successful manner can be judged by the present day development of man. In terms of number, technological and scientific capability we are still displaying the adaptability to changing circumstances which were so characteristic of our earliest forefathers.

One major difference exists, however. Modern man rarely allows himself to be involuntarily subjected to the changes of the natural environment. As a species, we now create change in our own right - economic, political, industrial and communication change - and it is the response to these changes which now dictates the lives of so many of our species.

Impact of Environment on Early Man

It is difficult to reconstruct the detail which surrounded the lives of early man. We have only the most infrequent archaeological sites to help in this task. It is also very difficult for twentieth century man to reconstruct the relationship which existed between early man and his environmental circumstances. Modern man is far removed from direct contact with the environment. Our comfortable homes are artificially heated and illuminated, we no longer hunt for food, we are accustomed to fast, convenient communication systems while the receipt of knowledge has led to a removal of the 'unknown' and a reduction in the associated emotion of fear.

But we must try and reconstruct the ways in which early man responded to his environmental surroundings because some of the early anthropogenic responses to environment have remained part of our character and have conditioned the way in which we behave towards our contemporary environment.

One misconception about our ancestors should be dispelled at the outset. Early man should not be called 'primitive' man. Early man was probably more innovative, more responsive and aware of his surroundings than we are today. By a process of experimentation he differentiated between plants that were poisonous and those which were edible. He discovered how to find water in arid areas, he developed means of stalking, killing and eventually cooking large, fleet-footed mammals. He would be capable of long-distance navigation, he could identify 'safe' resting places; he could, in effect, do far more than simply survive on the face of the earth - he could thrive.

Despite early man's great strength of character he also had many weaknesses. His great sense of fear was entirely an animalistic emotion. Gradually, he compensated for his instinctive fears by constructing around him the trappings of a civilisation which included totems, taboos and the rudiments of religion (Harris, 1971). Those great 'unknown' areas were transmitted into the earliest art forms - the rock paintings which show lightning and storms, the constant repetition of the saga of hunting, Fig 3.2, and geometric patterns which are thought to be attempts at understanding astronomy all came to bear upon the emergence of a human civilisation.

Fig 3.2 Typical Early Rock Painting. Deer Hunt from Los Caballos Shelter, Valtorta Gorge, Spain from Christensen (1955)

The original location of man has been traced with some certainty to the high veldt plateau of southern Africa (Clark, 1959). From this focal point man migrated northward through the African continent with divergent branches moving into the Mediterranean Basin and another heading eastward through India into China. Migration over distances of many thousands of kilometres would have exposed our ancestors to a vast diversity of environments. The responses to the different environments have been inherited by all contemporary human beings. The fact that even early, non-technological man could more than survive in a series of diverse environments has bestowed upon us a sense of superiority towards the environment. Our inherited attitudes tell us that our biosphere can be used to our advantage (Darlington, 1969).

First Attempts at Biosphere Management

The earliest members of our species made very little impact upon the ecosystems with which they came in contact. As Nicholson (1971) so aptly remarked, "the capacity to use and create fire [allowed] the mischief-making capability of the species to become manifest."

The use of fire as a technique which could be controlled by man represented the first step in a long and continuous chain of events which allowed our species to manage the resources of the biosphere to suit our own requirements (Stewart, 1956). Prior to the use of fire man's ability to modify the biosphere was insignificant but the advent of fire allowed widespread change to occur. Open vegetation (scrub and semi-desert areas) would burn easily while even dense forest with its tangle of undergrowth can be vulnerable to destruction during times of drought (Phillips, 1974).

Fire was probably first used to frighten marauding animals from camp sites. Later, fire was used to harden digging sticks, to cook animal flesh and later still was used as a means of driving prey towards a killing point. Simmons (1979) has indicated that the use of fire as an indiscriminate means of ecosystem destruction probably first occurred some 700,000 years ± 10,000 years before present (BP). There is no means of determining if the first signs of fire (from charcoal remains) were the result of chance, lightning fires or were deliberate man-made fires.

The most significant feature of fire upon the environment was its partiality to affect some components of the biosphere far more than others. This selectivity which is attached to the use of fire as a management technique allowed man to introduce, albeit involuntarily, a powerful means of directing the future evolution of some ecosystems. Fire was of particular significance in determining the development of low latitude ecosystems, for example the tropical forests and the savannas. Mid-latitude, temperate forests were also subjected to repeated burning. Plant and animal species of these ecosystems showed very varying response to the fire. The slow moving animals would be burnt, as would many of the small, delicate shrubs and herbs. Animals which possessed the ability to move rapidly and with the necessary stamina to keep moving for several days ahead of the fire would survive. The gradual replacement of forest by grassland allowed an increase in the number of grazing ungulates (e.g. deer). Mellars (1976) has suggested that herbivores may have increased in numbers by between 300-700% following the replacement of woodland by grassland.

Fire undoubtedly produced the first surge of anthropogenically-motivated species mortality. Slow-moving animals and small, delicate plants would have been exterminated. Plants with deep roots and tall stems with thick outer cuticles would have survived. These are the so-called pyrophytes.

The pre-burning ecosystem balance would have changed. Competitive forces between adjacent ecosystems would be removed and/or realigned. Former ecosystems in which all living space had been filled to produce 'closed ecosytems' would now have open patches suitable for in-coming colonising species. Gradually, as man's ability to work with fire improved more difficult sites were burnt. Jacobi et al (1976) has recorded Mesolithic man in Britain using fire to clear woodland in damp upland sites. Stewart (1956), in a classic review of the use of fire has shown its use rapidly spread to include wet rainforest sites, swamps and estuaries.

Agriculture as an Ecosystem Management Technique

The replacement of a hunting-gathering society by a settled agricultural community in which orderly cultivation and tending of specific plant and animal species predominated was one of monumental significance for mankind. It was achieved by a complex and reciprocal set of interactions; these interactions included major technical achievements, for example building of pens or stalls with which to retain animals. The emergence of agriculture also demanded a maturation of the human mind. In a conceptual sense, the realisation by early man that a new plant could be obtained by sowing a small seed must surely have been an astonishing and an incredibly difficult concept to accept. Watts, (1971) has suggested that the first stages in cultivation of plants would have been an accidental and erratic process. He suggests that a likely location for cultivation to have begun would be the midden piles which must have been a feature of all early settlements. On the mounds of rotting debris some seeds would have germinated and grown quickly in the nitrogen-rich conditions. These species would have been the nitrophiles; they would have attracted the attention of man because of their large size, their green, shiny leaves and perhaps also because of their soft touch.

By a process of trial and error some plants were found to be edible, some poisonous. Through the process of hybridisation some of these early agricultural plants became more adapted for cultivation by man (Bayliss-Smith, 1982). The true 'cultigens' gradually emerged, not as part of a deliberate policy by the early farmers, but due entirely to natural selection working in response to new and especially favoured environments which had unconsciously been made by man (Baker, 1970).

From evidence presented by Lee (1968), the early agriculturalists of the inter-tropical zone made greatest use of plants and only a little use of animals for their food supply. In mid-latitudes both plants and animals provided a food source, while beyond latitude 50° north animal food sources predominated. This pattern is not unexpected as the extreme seasonality of northern latitudes made plant growth a hazardous enterprise for the early farmer.

Agriculture was well established some 7000 years ago in a number of widely separated locations. Thus at Jarmo (in present day Iraq) a flourishing agricultural system has been identified (Harris, 1972). Wheat (einkorn, Triticum monococcum) and naked, six row barley (Hordeum vulgare) was cultivated. Evidence of agricultural development in many of the great river valleys of the Far East, the lower Nile Valley and, chronologically a little later, from the river valleys of Central and South America has been discovered.

The multiplicity of agricultural sites prompted Sauer (1952) to postulate his theories of "plural origin" in which he suggested separate, spontaneous origins for agriculture. It is important to remember that settled agricultural systems are relatively recent features of man's development. In the context of this book, the advent of agricultural systems mark an important datum point in man's development. From about 9000 years B.P. it became possible for agriculture to provide a constant source of food from a finite area of land. It produced a division of labour amongst the sexes, it allowed some of the populace to develop skills in other areas (weaving, pottery, metal working). It marks the point at which mankind assumed a conscious, role-playing part in the evolution of the landscape of this planet. Whereas fire had made the initial clearance of natural ecosystems a possibility, a variety of different agriculture systems under the direct control of man would cause massive and deliberate removal of ecosystems and species.

Emergence of agricultural systems permitted man to cease his previous nomadic life style and to replace it by one of relative stability. Certainly, seasonal transhumance persisted until relatively recently in agricultural communities (Egli, 1978) but the movement was relatively small scale and was between permanent bases, for example between a fixed winter settlement in a sheltered site and a more temporary summer location often upslope from the first settlement.

A relatively settled human population had serious implications for non-domesticated plants and animals. Any plant or animal species which interfered with the well-being of the agricultural system would be classified as a competitor (a weed or a pest species) and efforts would be made to exterminate the competitor. The early farmer was still severely limited in the land he could cultivate. Pockets of cleared land would have existed amongst a predominance of untamed 'wilderness'. But as man was highly selective of the land he could use, being confined to light, sandy, well drained soils, it is reasonable to hypothesise that specific ecosystems would have been put under threat of extinction. As his technological capabilities improved then so more sites were cleared for agriculture, isolated plots became continuous, wilderness areas became islands until eventually a complete transformation of the landscape would have occurred.

Jarman (1972) has argued that our ancestors

developed a survival strategy in which stable relationships between plants, animals and man were selected in favour of short-lasting, spectacular relationships. In effect, man was behaving exactly as any other component of the biosphere in which constancy of performance is deemed to be of greater advantage than maximisation of output.

An optimum, as opposed to a maximum level of food production, was assisted by the prevailing circumstances. These may be summarised as follows:

1. The population size was increasing only very slowly.

2. A diverse and numerous population of wild animals (in particular, mammals) provided an abundant source of meat.

3. Agricultural systems showed wide regional variation and were based on indigenous species.

4. The climate of the time (7000-3000 yrs BP) was both stable and favourable (the so-called 'climatic optimum' of the post-glacial).

5. Soils were relatively unaffected both from erosion and leaching and were attaining their maximum post-glacial maturity.

It is probable that during the lengthy time period when agricultural systems were being perfected, the early farmer still practised his hunting skills in order to supplement his diet, Fig 3.3. Archaeological excavation of Mesolithic sites in Europe supports this idea as large deposits of wild animal bones have been revealed (Jarman op cit).

It is relevant to note that apart from the elk, bison, bear and wolf, consumption of the animals shown in Fig 3.3 did not lead to their extinction. Animal population sizes were substantially reduced. Whether this was due to the direct effects of hunting or more to an indirect reduction in numbers through removal of habitats is hard to tell. In all probability the species listed in Fig 3.3 represent those which were capable of living alongside man and his agricultural land use. The first major period of animal extinctions probably occurred prior to 9000 years BP and involved species which could not tolerate even slight interference from man. Details about these species is still very vague.

Fig 3.3 Identified Animal Bones from 165 Mesolithic Sites in Europe

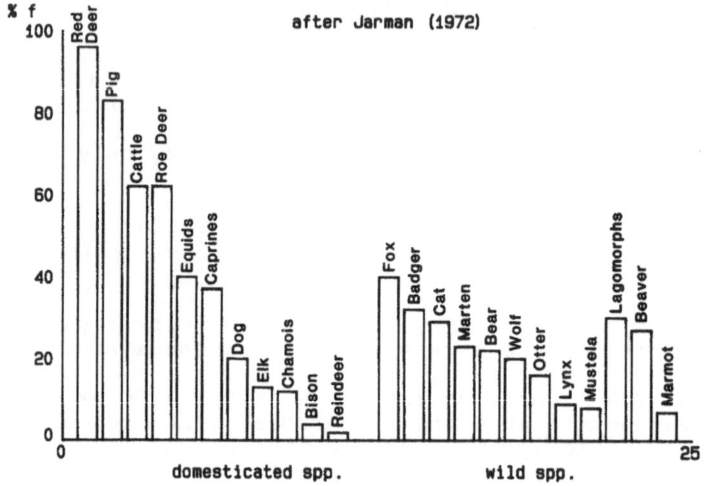

Continental Variations

Circumstances in North America were very different from those in Europe. In the latter, the arrival of man had been gradual and had occurred over a long time span - perhaps as much as 20×10^4 years. When man first arrived in North America (about 12,000 years BP) it was as a skilled hunter, armed with simple tools and with the powerful weapon of fire at his disposal. Point of entry was probably via the Bering Strait into present day Alaska. The time of man's arrival is no coincidence. Some 12,000 years before present the Pleistocene Ice Age was in its final throes. A land (or ice) bridge may have existed where now lies the Bering Strait. The people who crossed this bridge were the descendants of proto-man who had originated 45×10^4 years previously in southern Africa. The circumstances in which they now found themselves, forced them to place total reliance on wild animals for their food supply. Climatic restraints ensured that vegetation productivity was low and the little phytomass which was produced was of a form unsuited for consumption by man (Jones, 1979).

Fortunately for the early arrivals to North America, there existed an abundance of large mammals and these appear to have formed the main source of food (Martin, 1976). Species extinctions of these large mammals occurred rapidly. Only in part was this toll due to the impact of hunting, for throughout this period (12,000-9,000yrs BP) of post-glacial history there occurred a very rapid climatic and vegetational change. Due to a combination of these two facts, Martin (op cit) has postulated that two-thirds of the large mammals which had existed at the end of the Pleistocene disappeared within a time span possibly as short as 500 years. This rapid extinction has been called the 'Pleistocene overkill'.

North America was an empty continent and the new hunters swept rapidly southwards reaching the Gulf of Mexico within 350 years of their arrival in the north-west. Reed (1970) has estimated that 24 genera of animals disappeared from N. America at the end of the Pleistocene and this compares with the loss of nine genera from northern Eurasia. It is worth noting that in the high latitudes of the Old World, species extinction was probably due entirely to post-glacial climatic amelioration as man had not yet penetrated the northern wastes. As Simmons (1979) has explained, large mammals were disappearing from both N. America and Eurasia but in the former the species extinction was greater because of the hunting losses due to man.

Ecosystem Changes Between 1AD and 1500AD

The previous section has shown that agricultural systems had developed in separate centres by 7000 years B.P. All the main agricultural crops had been developed (wheat, oats, rice, beans, proto-pea) as well as the domesticated animals (dog, goat, sheep, ass, horse, cow). Agricultural land use was still new, there were fewer people following a settled agricultural-based life style than the majority who were still hunters or primitive pastoralists. The total world population at 7000 BP has been estimated by Deevey (1960) to have been approximately 78×10^6. Very gradually, agriculture became capable of supporting an ever increasing population. By 1500 AD estimated world population was 187×10^6, Fig. 1.1b.

An inevitable consequence of this trend towards organised agriculture was the inevitable disappearance of the natural vegetation. Gradually, the natural ecosystems of those areas below about 300 metres above sea-level were replaced by an 'organised' agricultural landscape in which fields, trackways and buildings replaced forest, marsh and small grassland patches.

Early Technology

What impact did early technology have upon the natural ecosystems of those times? The harnessing of two or more oxen to a simple, wooden plough enabled man to scarify the soil. The original plough was unable to turn over the earth - it merely scratched a shallow furrow in the surface. This 'machine' was capable only of tilling the lightest soils from which vegetation had been removed by some other means (fire, grazing or by man's own labour). This simple plough could not bring the heavier, wet soils of northern Europe into cultivation but by 700 AD a new, advanced plough had been developed. The new plough sliced and turned over a single furrow of land - it could cultivate clay-rich and wet soils. There was, as ever, a price to pay for technological development. The tougher soils demanded a team of 6 or 8 oxen which in turn had to be fed and watered. To grow the necessary fodder to feed the extra oxen, more land had to be brought under the plough. Few farmers could at first achieve this extra demand. Prior to 700 AD farming communities were self-sufficient, subsistence units.

The emergence of the new plough changed man's relationship with the land which supported him. Whereas the farmer had previously cultivated only enough land to support the food requirements of his family, in the changed circumstances the cultivated area was dependent upon the capabilities of the machinery available to till the ground. White (1967) has argued that it was this development, and not the earlier fire-based technologies, that enabled man to detach himself from the constraints of living within the confines of natural ecosystems. Once free from the limitations imposed from within the ecosystem, mankind rarely paused to look back upon the void his departure had made within the ecosystem. Freedom from the constraints of ecosystem boundaries made mankind forget about his obligations towards the other components of the ecosystems on which he was still dependent. Instead of cooperating, man now embarked upon drastic exploitation of the ecosystem and biosphere components. Habitat destruction proceeded apace and along with it occurred major species extinction.

Man's technological skills were not focused solely on his methods of food production. One of the great characteristics of Homo sapiens was that new skills were grafted on to existing levels of development. No skills were lost, merely reallocated in rank order. Thus, although man soon became a sedentary agriculturalist he still supplemented this life style with occasional hunting forays. The migratory senses of early man had not been lost. Exploration and discovery took man into every part of the planet. Even today that trend has not stopped. Travel, if only in the form of an annual summer holiday, satisfies our need to explore our planet, while at an extreme level, the current attempts to travel to other planets in our solar system is an example of mankind's insatiable curiosity and innovative skills.

Expansion of Man's Technology

By about 1500AD man was technologically and mentally confident in his capabilities to conquer the planet. Any species, plant or animal, which interfered with man's well being was eliminated, or else forced to take up a location at the periphery of man's own territory.

A feature of agriculture in the period currently under consideration was the need to allow a fallow period when the land was allowed to 'rest'. The fallow time, coupled with the small size of fields or strips of cultivation, allowed the still-wild plants and animals to migrate from one area of land to another. Flannery (1969) has suggested that if these wild species became too numerous then their populations would have been deliberately reduced through hunting or weeding. Biological diversity was such, however, that if a species was deliberately removed then its vacated 'niche space' would have been quickly filled by another species. The grasses of the genera Avena, Cynodon and Phalaris illustrate the process of cultivated species replacing wild grasses.

It is inevitable that during this phase from the

rise of Christianity to the Middle Ages that many plant and animal species would have been exterminated through direct persecution. Very many others would have had their distributions severely curtailed as habitat alteration, consequent upon the spread of agriculture, brought about a shrinkage of natural habitats. Fortunately, these changes brought about by man were introduced relatively slowly and were still small-scale in their geographical extent. Species were able to adapt to the changing conditions. This was particularly so for the still wild animals which lived on the edge of man's habitat. Many species of small birds quickly learnt that man's habitation provided a ready source of food and nesting sites. Bats, mice, rats, fleas, ticks, spiders and beetles also adapted their life styles to accommodate man's presence.

Ecosystem Changes Between 1500-1900 AD

The beginning of this time period, circa 1500AD, was marked by a still relatively small human population - probably a little below 187×10^6 and of which only 10% lived in nucleated settlements. Technological developments were still minimal, and confined in a geographical extent to the Mediterranean Basin and central Europe. Water wheels, windmills, a plough with a moldboard and weight driven mechanical clocks represented the state of the technological art.

In 1543 two great scientific works were published, Fabrica and De revolutionibus by Copernicus and Vesalius respectively. These publications are generally thought to mark the beginning of modern science (White, 1967). They mark a new awareness on behalf of man of his surroundings and in turn they herald a new assault upon ecosystem and species survival. By 1650 the latent biological productivity of the human animal was beginning to show signs of a metamorphosis. Durand (1967) has extolled the need for caution in interpreting the scanty evidence to suggest a population increase in the century between 1650 and 1750, though he admits that accelerated growth was probably occurring in parts of Europe, Russia, China and to a lesser extent, in N. America. By 1750 a marked increase in the human population curve was evident, Fig 1.1b. It has been assumed that the catalyst for this increase was the improved economic climate created by the Industrial Revolution which was underway in the older European nations. In a detailed analysis of world population figures Durand (*op cit*) has suggested that several non-industrial nations, for example Russia, were exhibiting a rate of growth equal to that of the newly industrialising nations of Britain and France. Similarly, the growth of population in China in the 1700s appears to be totally unconnected with industrialisation.

Transfer of Exotic Species

One possible explanation for the growth in world population during this period can be found in the broadening of the agricultural base through the transfer of species between the Old and New Worlds. Plants such as maize, tomato and cotton were brought from central America to Europe, millet from Africa, onion, peach and brassica (cabbages) arrived from the Far East as cargo on the returning boats of the great explorers (Hylander *et al*, 1941). Many of the introduced species were collected for their ornamental or novelty value and their economic value was discovered accidentally at a later date (Whittle, 1975). This is certainly true for the introduced tree species such as Japanese larch (*Larix leptolepsis*) and the Douglas fir (*Pseudotsuga taxifolia*).

The impact of an 'exotic' species upon a native flora and fauna can have the gravest consequences. Elton (1958) has described some highly successful arrivals as 'invaders'. Thus, the rabbit, introduced to the State of Victoria in Australia in 1888, underwent a population explosion. Likewise the common sparrow when introduced to New York State (Kormondy, 1969) and the mongoose (*Herpestes auropunctatus*) introduced to Jamaica in 1870 to control the native rat population (Simmons, 1979) both showed a catastrophic increase in population size due, in large, to the absence of native predators.

Native fauna and flora of New Zealand also suffered considerably from introductions. Captain Cook who visited the islands in 1769/70 put ashore a small population of domestic pigs, with the hope that they would increase in numbers so providing a supply point for future mariners who visited the islands. The pigs dispersed into the forest destroying as they went the nests of many of the unique ground dwelling native birds of New Zealand (Kuschel, 1975). Also from the antipodean islands come the examples of the so-called noxious weeds - bramble (Rubus spp.), gorse (Ulex europaeus) and bracken (Pteridium aquilinum), all of which have been successfully transferred from the northern hemisphere.

Extinctions

While the transfer of 'new' species can be recorded with comparative ease, for example from explorers' log books and from naturalists' field reports, the disappearance of species usually goes unnoticed. Only the largest extinctions have chance of being recorded. Thus, the ancestor of the domestic cow, the auroch (Bos primigenius) is known to have finally succumbed to over-zealous hunting in Poland in 1627; by about 1790 the beaver (Castor fiber) had been hunted to extinction in central Europe, and by 1743 the wolf had been exterminated in Britain (Ziswiler, 1971, Simmons, 1979). In these three examples we find the most common reasons for the extinctions of animal species; the auroch provided good sport, the beaver was a 'multi-causal' extinction - its pelt was valuable, an extract from its body was believed to have anti-arthritic properties and it was of nuisance value due to its propensity to dam rivers. The wolf was exterminated solely because of the damage caused to man's domesticated animals and due to the fear of wolf packs attacking man himself. Fisher et al (1969) have claimed that of all known extinctions which have occurred since 1600 AD, three quarters of the mammal extinctions and two-thirds of avian extinctions have been directly due to man, the remaining causes of extinctions being due to natural causes.

Hunting by man is the single greatest cause of extinction, 42% of avian extinctions and 33% of mammalian extinctions being attributable to this cause. Reference to Fig 6.7 suggests that between 1650 and 1850 species extinction was comparatively rare being less than five taxa of mammals and birds per century. This figure should undoubtedly be much higher, as many species would have passed into oblivion without ever having been given a name by man!

In addition to direct extinctions through hunting, man has increasingly been responsible for extinction through introduction of aggressive 'exotic' species and, in particular, from habitat destruction. It was shown in Chapter Two that every species of plant and animal must possess a habitat of sufficient size and productivity in order to ensure survival. The progressive clearance of natural vegetation to make way for agricultural land, the elimination of 'waste' ground, the draining of swamps and removal of woodland have all caused mass alteration of habitats. Ploughing of land altered the soil environment and disrupted the soil moisture conditions. During the period 1500-1900 AD these changes became more extensive and occurred faster than ever before. Adaptation to these changes became more difficult for plants and animals to achieve (Guggisberg, 1970) and this is reflected in the considerable rise in the rate of extinction post 1850 as shown in Fig 6.7.

Ecosystem Change in the Twentieth Century

The trends which had been established in the previous period of history have continued, unabated, throughout the twentieth century. Thus, introductions, hunting, predation and destruction of habitats proceeded apace. Several additional factors have also occurred, the effects of which have been to substantially increase the rate of species extinctions. Those additions are:

a) Pollution of the biosphere (the land, sea and air).

b) An intensification of the agricultural system as indicated by:

1. Replacement of fallow by continuous cropping.
2. Replacement of indigenous crops and stock by complex hybrids which have been genetically engineered.
3. Control of pests, disease and weeds by chemical means, e.g. use of pesticides, insecticides, herbicides.

c) A major increase in the use of complex, chemical compounds many of which are nonbiodegradable.

d) An exponential increase in the human population with an additional demand created by greater mobility, greater affluence and greater expectations of personal ambition and achievement.

e) Introduction of maximisation methods of productivity - factory farming, factory fishing fleets.

The combined impact of both traditional and new hazards on ecosystems and species has been substantially greater than that which had occurred previously. Even allowing for the greater concern and the more accurate measurement of species loss over the last twenty five years, the impact of contemporary man upon the biosphere and its resources has been catastrophic. Species which are currently being made extinct are considered in greater detail in Chapter Six. It is sufficient here merely to indicate that the IUCN estimate that one species of plant and/or animal may be lost each day between 1975 and 2000AD (Allen, 1980). For vertebrates alone, the extinction rate is currently about one species or sub-species per year compared with a rate of one species per ten years in the period 1600-1950. Traditionally, we assume that the period of greatest animal species extinction was during the cataclysmic episodes which occurred during geological eras of the past. The extinction of the dinosaur group during the Cretaceous era was a notable example of large animal extinctions but based upon an accurate fossil record, palaeontologists have established a dinosaur extinction rate of one per 1000 years (Myers, 1976).

Man the Modern Hunter

Man has now attained a complete technological dominance over all other living creatures with the possible exception of virus and bacteria groups. For example, hunting is now an expensive pastime for the rich financial or industrial technocrats. The chase, often the longest and greatest challenge to the skilled hunter, has been replaced by a short, efficient helicopter flight from which the kill is made with high-powered, repeater rifles. The dead animals are collected by a back-up helicopter and taken to a deep freeze factory for storage. Meantime the 'hunter' relaxes over his drink alongside the swimming pool at the luxurious hunting lodge. This scenario is not fiction but is a commonplace event throughout North America, parts of eastern and southern Africa and in New Zealand. It is impossible to implement a policy of selective culling of old and infirm animals when hunting becomes a high cost, high speed international business and it is little wonder that the age and sex ratios of many herds of wild herbivores is such that the stability and hierarchical structure of the animal community is thrown into disarray.

Destruction of habitats as a result of man's actions is yet another major problem faced by many plants and animals. Often, the replacement of some habitat by another will be rapid and total, for example the construction of houses or a factory on a greenfield site or the replacement of a natural forest with a managed forest comprising exotic species. On these occasions the destruction of the original ecosystem will be obvious. The loss of species will be easily monitored by 'before' and 'after' species surveys.

Far more difficult to assess are the slow, often imperceptible changes which are taking place in ecosystems, partly as a result of man's actions and partly due to natural environmental change. A good example of this type of change can be found to occur in the distribution and composition of the Karoo vegetation in South Africa (Acocks, 1975). At first glance the Karoo is a monotonous and changeless ecosystem in which moisture availability is the most relevant controlling factor. Closer inspection shows the Karoo to comprise a mozaic of different sub-units which have evolved as a result of micro-variations in moisture availability, intensity and duration of frost, grazing pressures, fire, soil type and soil erosion (Meadows, 1985).

Fig. 3.4 shows the extent to which the Karoo ecosystem has already increased its distribution and also predicts the extent of future advancement of this ecosystem. Field records made by Acocks (op cit) since the early 1950s have shown that in the vicinity of Kimberley, Karoo communities have replaced grass species of the Thornveldt and in the southern Orange Free State, Dry Cymbopogon - Themeda Veldt has been overrun by Karoo species. The distribution of vegetation communities is never static. In Southern Africa vegetation is always trying to migrate into drier habitats. It is unfortunate that the Karoo ecosystem represents an extension of less productive plant species which in turn can support fewer grazing animals. The South African farmer, in his attempts to maintain his farm productivity and income levels, will maximise the use of the grazing potential but by so doing will only increase the rate at which the Veldt type (grass dominated) ecosystems will degrade towards Karoo type (open bunch-grass/thorn scrub) ecosystems.

Fig. 3.4 Advancement of Arid 'Karoo-type' Ecosystem, South Africa.

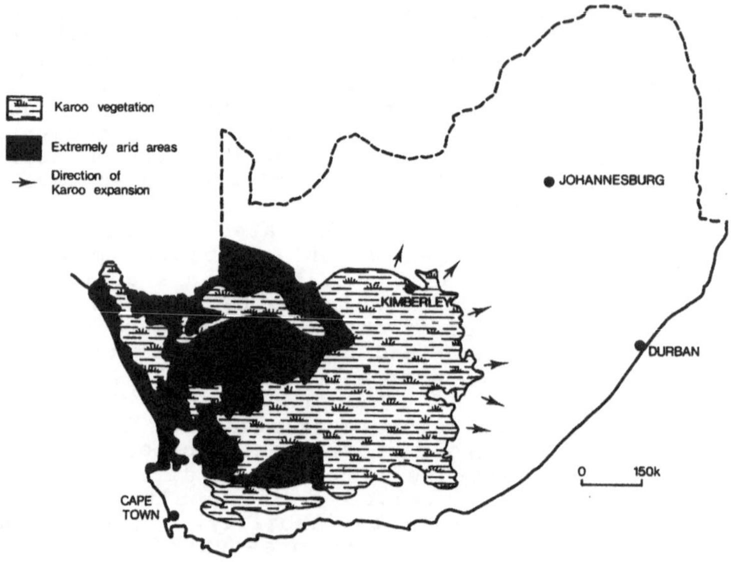

'Improvements' to the Biosphere

One of the most persistent features of mankind's behaviour is the desire to 'improve' both upon nature's design of plants and animals and the relationship of organisms with their environments. Domestication of plants and animals was the very first way in which we directed the evolution of natural species towards a 'design' more suited to our own requirements. The prime requirement was that of a guaranteed food supply. The story of agriculture over the ages has been one of increasing intensification of production through the ever closer control of the biology of the plants and animals within the agricultural system (Beresford, 1975).

The intensification of agriculture, particularly in the last 20 years, has resulted in many indigenous plant and animal species being unable to locate themselves in suitable life zones. As a result they have been pushed towards their zone of tolerance (Cox et al, 1976). Even on the occasions when native species are successful in retaining a niche space the vigour with which they can complete their life cycle is often retarded due to the operation of the modern agricultural system. The 'hazards' faced by native species are shown in Table 3.2. The modern agricultural system can only be successful if the farmer is capable of satisfying most of the requirements given in the left hand column of Table 3.2. Munton (1974) correctly states that intensive farming discourages the multiple use of farmland. The competitive land users - in this case native plants and animals - are either actively removed (hedgerow removal and killing of foxes, badgers) or conditions are so unsuitable for their survival that they are forced to migrate. Figure 3.5 shows that in the early stages of agricultural development (time t1), there are abundant areas in which native species can complete their life cycle. In the condition shown in time t3 the unused areas have shrunk to small sectors of land. If these unused areas are separated from other wild areas then the native populations become trapped by the unfavourable, managed ecosystems which surround them. Only when corridors are allowed to exist between the unused areas can species migration occur.

The concept of habitat corridors has been developed in North America (Wintsch, 1986), in particular in the southern states of Georgia and Florida where plans exist to link the Okefenokee National Refuge and Osceola National Forest by habitat corridors, with possible future extensions to the coastal refuge sites, Fig 3.6.

Society in general, and farmers in particular, are increasingly recognising the impact of contemporary land use changes upon native plant and animal populations. The ecological and conservationist-based arguments for protection of native species are now increasingly accepted alongside the more traditional economic arguments. A major dilemma faces agriculturalists, conservationists and politicians. To what extent can a conservation programme be incorporated into modern farming practice?

Fig 3.5 Advance of Agricultural Land and Decline of Wilderness Areas

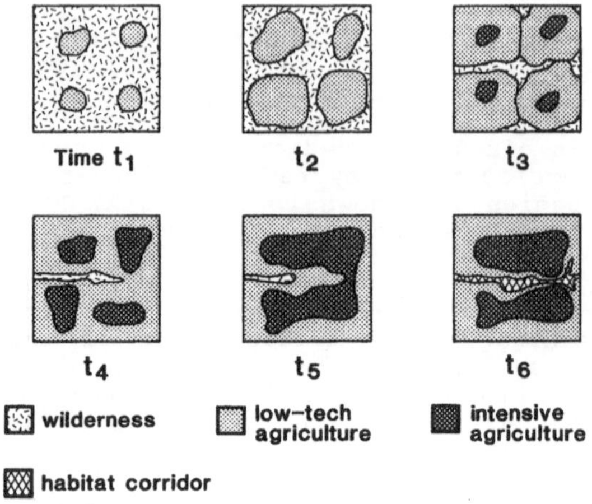

Fig 3.6 Habitat Corridors in Georgia and Florida after Wintsch (1986)

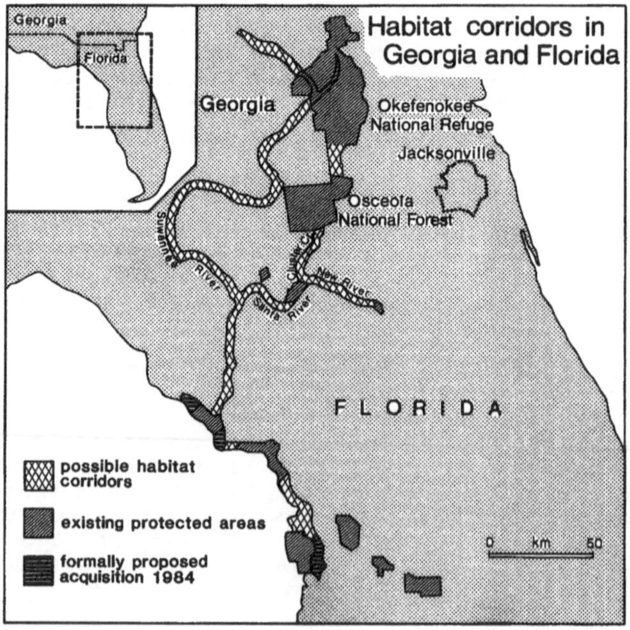

TABLE 3.2 Characteristics of Modern Agricultural Systems and the Impact on Native Plants and Animals.

Agricultural system	Consequence for Natural System
Large field size	Removal of hedgerows and shade trees Reduction of habitat 'refuge' sites
Optimum soil moisture levels (drainage/ irrigation)	Disappearance of 'boggy corners'. Uniformity of soil moisture levels produces a standardised soil environment. Reduction in soil habitat diversity levels.
Specialisation on a few crops and/or animals	Monoculture produces a visually monotonous landscape. Enables outbreaks of disease to spread rapidly if uncontrolled. Reduces habitat diversity for native species.
Extensive mechanisation	Enables farmer to complete soil preparation, planting and harvesting more quickly. Prevents animal species from completing their life cycle due to higher 'disturbance' factor. Mechanisation can lead to increased local pollution level, eg diesel fumes and chemical spillage.
Use of hybrid crops and selectively bred animals	All modern agricultural crop and animal species have a significant proportion of 'foreign' (exotic) genetic material. These species demand very high levels of management inputs to ensure optimum conditions for growth Inter-breeding or crossing with native species not possible, thus a formerly important avenue of evolution prevented.
Use of inorganic chemicals (pesticides, herbicides, insecticides, chemical fertilisers	Monoculture and hybridised species demand high feeding levels. Large fertiliser applications can contaminate wild life - accumulation of chemicals in food web. Reduces breeding success of native species

The farmer is committed to producing food as cheaply and effectively as possible and yet at the same time would be willing to consider the setting aside of conserved areas and habitat corridors provided that financial compensation was available to offset income from lost production. Reference to Fig 3.5, time t6 shows how habitat corridors might be integrated into an agricultural landscape. Harris (1984) has advocated that landowners who participate in this form of conservation should be allowed substantial reductions in property taxes. There is little doubt that the creation of habitat corridors and conserved ground will be a charge which the general public will have to pay for through some form of taxation (a direct conservation levy or an indirect tax in the form of increased payment for food). The EEC has made £6 million available in 1986 for farmers in six areas of England and Wales to return marginal areas of farmland to so-called 'traditional farming' methods. The money will offset losses in production through reduced usage of chemical sprays, replanting of hedgerows, repair of stone walls and general encouragement of wild life species.

While welcoming this approach it should be noted that it applies only to a tiny fragment of the total agricultural area of England and Wales. Conservation activist groups have criticised the scheme as satisfying the conscience of Government yet at the same time goes little way towards satisfying the requirement for conservation in agricultural areas.

Recent Trends in Ecosystem Change

The deliberate removal of species by hunting and eradication programmes has an obvious deleterious effect on the survivorship chances of many plants and animals. Removal of a single species will weaken the total food web of which it forms a part. If several species are removed from the same food web, and particularly if those species are located in the same trophic level, then the entire feeding structure can be thrown into jeopardy, see Fig 2.5.

In addition to the direct, and often deliberate removal of species by man, must be added the unintentional losses of species which can come about by the intensity of use which mankind now places upon some ecosystems. Also, in our attempts to place greater management controls on ecosystems we make increasing use of chemical substances to boost ecosystem productivity (eg. agricultural systems boosted by inorganic fertilisers), or chemicals to control unwanted pests and diseases (e.g. dichlorophen to control lichen and algal blooms). Some of these chemicals become resident in our biosphere and accumulate until they reach toxic levels. These form another vast area of concern for conservationists. Man has always been responsible for creating pollution of the biosphere, but as with so many environmental issues, it is only in the last 25 years or so that we have come to realise the dangers inherent in a polluted biosphere (Holdgate, 1979).

Pollution of the Biosphere

Innumerable numbers and sources of pollutants now exist which can jeopardise ecosystems. Environmental pollution has become a more serious problem as the human population has increased in size and as industrial processes release more complex forms of pollutants.

For countless centuries, pollution was not recognised as an environmental problem. Certainly, in Britain during the Industrial Revolution and into the Age of Steam (a period stretching approximately from 1750 to 1914) the well known dictum of 'where there's muck there's money' went virtually unchallenged. Pollution of land, air and water was an acceptable corollary of industrialisation and it is difficult to understand why the somewhat puritanical, Victorian society which gave the illusion of believing that 'cleanliness was next to Godliness' were so prepared to tolerate such gross despoilment of their environment.

The improvement in the life style of ordinary human beings which was achieved by the Industrial Revolution overshadowed the degradation which occurred in many parts of the natural environment. Many of the ecological casualties which resulted from the nineteenth and twentieth century technological revolutions could not be predicted, nor identified because never before had such immense and rapid change to man's life style been achieved (Ford, 1970).

Man's Changing Awareness

An important change in the maturity of western civilisation occurred over the decade 1965 to 1975. For the first time the public was made aware that mankind was capable of making large scale changes to our planet. A number of American researchers put forward papers and books in which the destruction of ecosystems and species was graphically reported. Hardin (1968), Commoner (1969), Detwyler (1971), Murdoch (1971) and Ehrlich (1977) all produced well articulated arguments and case studies of the ways in which mankind was threatening the stability of the biosphere. Television documentaries transferred these academic arguments to the layman. Bronowski (1973) narrated a much publicised "Ascent of Man" in which the uniqueness of *Homo sapiens* as an animal species was clearly identified along with the many problems of his making.

Nowhere was this awareness more apparent than in man's newfound realisation that pollutants were destroying the quality of land, air and water. In turn, the loss of quality was reducing the vigour of many plants and animals. Stern (1977) edited a major five volume work on air pollutants and has devoted considerable attention to the impact of pollution on the biota. Marsh (1947) indicated clearly the effect of pollution on trees, describing the symptoms as:

> a stunting of growth, loss of vigour, reduction in reproductivity capacity, degradation of colour and ultimately death.

Woodwell (1970) described how the accumulation of pollutants occurred on the time scale of succession and not on the time scale of evolution. As a result, species have little time to adapt to the degraded environments. Lower light intensities and differing light qualities upset the balance of photosynthesis in some plants. Specialised species gave way to the generalist species, diversity in ecosystems is replaced by monotony, stability in ecosystems by instability, and our biosphere

> moves away from a world that runs itself through a self-augmentive, slowly moving evolution, to one that requires constant tinkering to patch it up, a tinkering that is malignant in that each act or repair generates a need for further repairs to

avert problems (Woodwell, 1970).

Mankind has come far in both literal and absolute terms since his emergence as a tentative species some $2.5x10^6$ years ago. We have shown ourselves to be capable of increasing our numbers from small, isolated family groups to mega-conglomerations, with a total population of $4.7x10^9$. We attain absolute mastery over most, if not all, non-human species of this planet, our technology has allowed us to selectively breed plants and animals for agricultural use, we have developed immense skills in medicine, metallurgy and aeronautics. The list is long and will inevitably become even more spectacular as we proceed into the twenty-first century.

But has the 'awareness' of man's ability to degrade and destroy the resources of our planet as well as utilise them to further our own ends been really appreciated by society, by economists, industrialists and politicians? In reality the situation varies from country to country. North Americans pay more by way of dollars and effort in safeguarding the ecosystems and species of that continent. In Europe the situation is patchy. Britain relies absolutely on strictly enforced planning legislation and tends to hide under the complacent cover provided by the legal protection.

Politicians and the Scientific Civil Service in Britain are slow to concede that an acid rain problem exists while their counterparts in Sweden maintain that sufficient evidence exists to show that acid rain not only occurs but that it creates major destruction of both the natural and built environments (Hutchinson *et al*, 1980; Shriner, 1982; Press *et al*, 1983).

Numerous attempts have been made to force attention onto the problems of biosphere degradation. In 1971 a new academic journal was published entitled *The Ecologist*. Part one of its second volume (*The Ecologist* Vol 2.1, 1972) made the headlines when it appeared under the title *A Blueprint for Survival*. The arguments made in the article were supported by 38 distinguished British scientists and its appearance brought about a heated debate among politicians, economists, industrialists and scientists.

A Blueprint for Survival made serious claims; for example it stated that governments throughout the world were either "refusing to face the relevant facts, or are briefing their scientists in such a way that their seriousness is played down". The claims made in the Blueprint are remarkably similar in their basic tenets to that made by the Rev Thomas Malthus in 1798 (Malthus, 1970) in that over use of resources would lead to shortages of such serious proportions that society, as it existed, would be unsupportable and a 'crash' would soon follow.

The warnings contained in A Blueprint for Survival were not specifically acted upon by any government. In that respect the Blueprint was a failure. The initial movement which accompanied its publications was not followed up. Many of the ideas contained in the document did, however, slowly creep into politicians' speeches.

Perhaps the greatest success of the Blueprint was the way in which it drew the attention of the public to the issue of conservation, viewed in its widest sense. It was paradoxical that soon after its publication the very foundations of our industrialised societies were plunged into confusion as a result of the energy crisis. In November 1972 the price of crude oil rose from less than $10 a barrel to a figure of about $16 a barrel. Inflation rates shot upwards. Oil prices continued to rise so that by the opening years of the 1980s a price of $35 a barrel on the spot markets at Rotterdam was commonplace. The catastrophic rise in the price of oil was only partly due to the acceptance that oil was a finite resource. The greed of the multinational oil companies combined with the avarice of taxation by the governments of producing nations was responsible for the greatest proportion of the price increase. Here is yet another example of man behaving true to his species characteristics - maximising the opportunity presented to him.

Chapter Four

DEVELOPMENT OF THE CONSERVATION ETHIC

Homo sapiens is the most successful animal species to have ever lived on this planet. While our physical frame prevents us from achieving *individual* feats which match other animals (we are not as strong as a horse, nor as fleet-footed as a cheetah, nor do we have the sight of a hawk nor the hearing of a gazelle) our physique is such that in a *compound* manner we are a very successful, adaptable, general purpose animal. We have one major advantage over all other animals; we walk on our hind limbs, thereby releasing our fore limbs for other functions (see page 30).

In order that we can gain complete control of our hands and arms, our cranial capacity has developed so that sensory receptors (touch, sight, hearing and balance) allow considerable dexterity over our movements (Bronowski, 1973). Unlike the lower animals, where movement is involuntary and is due to changes in the environment, the vertebrates have substantial control over their movement patterns. In man, the automated control mechanisms can be overridden by the deliberate use of powers of concentration, determination and by training. Danserau (1966) has called this behavioural trait the 'Law of Persistence'. Few other animal species have developed this ability. It has allowed our species the ability to make extraordinary physical and, more importantly, mental achievements. It has given us the conscious control of our day-to-day lives.

We have used our manual and mental dexterity to take advantage of our immediate surroundings. We have used biotic and abiotic components in the biosphere to support the infrastructure of our civilisations. We differ from other animals in the diversity and sheer volume of use of the biospheric resources.

Until the 1960s mankind treated his resource base as if it were a never-ending, infinite supply of materials. This was an inevitable behaviour pattern as until that time new resources were being discovered and new lands were still being opened up. Northern Canada and Alaska were explored for geological resources, the central Soviet land mass was brought, partially, into successful agricultural production while in tropical Africa and South America a communications network allowed human penetration into areas hitherto confined to native populations.

Our very success in using the biosphere resource base now threatens our continued well-being. If mankind was merely another animal then our progress as exploiters of the biosphere resource base would have been checked long ago by natural competitive forces. But man can no longer be considered as merely another member of the animal kingdom. Our early biological superiority in terms of dexterity and cranial capacity have now been translated into scientific and technological capabilities which allow us to tackle undreamt-of tasks.

Sears (1956) has claimed that our success is due not only to our evolutionary superiority but is due also to the highly evolved and specialised environments in which we find ourselves. The existence of an angiosperm flora (in particular, the presence of legumes and grasses) and a mammalian fauna which feeds upon this flora have combined to provide man with a readily utilisable resource base. Furthermore, Sears (op cit) has argued that it is the stability and orderliness of the biosphere in respect of energy and material movements which has provided the basis for the existence of such a complex organism as man.

It has been shown in Chapter Three that man developed in a gradual, progressive fashion supported by an increasing use of biosphere resources. These resources have, in part, comprised the physical components of our planet such as the soil, water and air and also comprise the biological components - the plants and animals.

The use of these resources has been deliberate; we require raw materials in order to survive. But we have gone well beyond the point of mere survival. We have used resources in ways which have allowed us to become increasingly sophisticated and, paradoxically, more dependent on the use of even greater quantities of raw materials.

While our biosphere shows signs of a natural increase in its inherent productiveness and diversity (Moran, et al, 1980), mankind has countered this trend by working towards a simplification of the biosphere structure. Sometimes we have acted deliberately; we have felled forests to make way for agricultural land and we have drained swamps to rid ourselves of disease-carrying flies. For every deliberate activity there are an innumerable series of events which occur as a 'knock-on' effect. For every hectare of forest we chop down we destroy the habitats of birds, small mammals, insects and a vast array of plant species, see Fig 2.5.

The Changing Role of Time

In the past, biospheric change has occurred over natural time scales. Thus, geological change is measurable on a time scale of tens of millions of years while biological and climatic change operates on a faster time scale of tens of thousands of years (West, 1968). For the bulk of our planet's history these changes have been entirely due to natural events.

From about 8000 BP Polunin and Huxley (1981) have suggested that the Mediterranean Basin came under the influence of a radically different type of change. For the first time man was responsible for large scale forest removal. Deliberate use of fire was followed by the grazing of sheep and goats which caused an eventual loss of top soil. These changes are collectively termed anthropogenic change. They now dominate the changes occurring in the biosphere. They differ in two ways from geological, biological and climatic change:

1. They are deliberate changes, brought about to enhance the survival of one species - mankind.

2. They occur on a much faster time scale than natural change - involving a time span of as

little as 25 years.

Initially it was assumed that as anthropogenic change was responsible for improving the lot of man then it must be beneficial change. This was, in all probability, an acceptable assumption when our early civilisations were making rapid and radical improvements in the quality of life style. More recently, however, short-term financial gains combined with the politician's four or five year term of office have produced a trend whereby the use of the planet's resource base has been mortgaged for the sake of immediate materialistic gain. It is impossible to assess the impact we have on the biosphere in as short a time as five or even ten years. To see the real impact of man on the biosphere we must work with a time span of about 25 to 30 years.

Two separate issues emerged in the 1960s and 1970s in which the impact of time had an interesting effect. The first of these was the growth of the human population which now appears set to double its size in about 33 years (ICP, 1984) and the second concerns the storage of nuclear waste materials. The nuclear wastes which we are currently producing will still be of dangerous intensities some 5000 years hence while plutonium 239 reaches its half-life only after 240,000 years (Report of the Nuclear Energy Policy Study Group, 1977).

These examples are different in that they represent man's involvement with two extreme time scales. The human population time problem is immediate in that our planet is already seriously overcrowded. Every day the situation becomes more critical because more people mean more demands being placed upon the biosphere. The nuclear power issue differs in that while it presents an imminent threat to the safety of the biosphere, the problem of nuclear waste disposal is a problem which will remain with our descendants for at least 1000 years into the future.

Growth of World Population

It was not until the first quarter of the present century had elapsed that politicians, medics, agriculturalists and indeed the general population of developed countries realised that the resource base upon which man lived was finite. Until that time the opposite had been suggested; that it was man's numbers that were finite. Malthus had argued that the human population would reach a peak beyond which it could not pass because of restrictions in food production (Sandbach, 1980). Animal ecologists and statisticians presented theories which also suggested that animal numbers behaved in ways which produced a finite population which fluctuated around the carrying capacity of the environment, Fig 4.1 (Clapham, 1973).

Fig 4.1 Habitat Carrying Capacity, 'J' and 'S' Shaped Population Curves

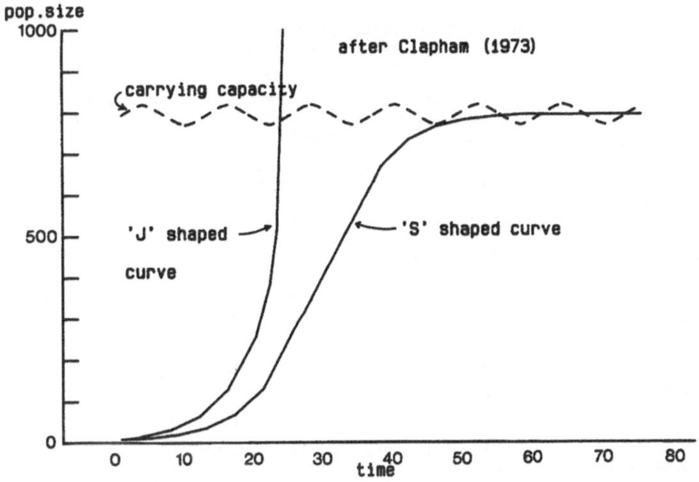

Reference to Fig 1.1b shows the pattern of human population growth from 10000 BP to the present with the predicted growth to the year 2000 AD. The pattern of growth follows that of an exponential curve in which less and less time is required to achieve a doubling in population size, Table 4.1.

Keyfitz (1971) has produced a thorough review of both the historic rise of population numbers and also provided predictions of future numbers. Working from United Nation statistics which suggested a rate of increase equivalent to 20 per 1000 per year, (+2.0% growth) and allowing for a gradual fall in fertility levels, the world population by 2000AD will have attained at least 6.0×10^9. Projections of population size are notoriously difficult to achieve with any degree of accuracy. It is not solely the total population size that must be determined but also data relating to child mortality, life expectancy and literacy levels. These three indices, when combined, give a good measure of the Physical Quality of Life Index (PQLI). Birth rates generally vary inversely to PQLI values though other variables such as health-care, employment levels, age of marriage, improved status for women and sexual equality of incomes can also influence birth rates and hence population growth (Myers, 1985).

Table 4.1 Growth of Human Population (1825-1985), figures from United Nations reports.

Population Size	Approximate Date Achieved	Time Required to Reach Population	Equivalent Doubling Time, (yrs.)
1.0×10^9	1825	1.0×10^6	-
2.0×10^9	1927	1.0×10^2	100
3.0×10^9	1966	33	70
4.0×10^9	1975	11	48
5.0×10^9	1985*	10	33
6.1×10^9	2000+	-	-
10.5×10^9	2100+	-	-

* 1985 figures are estimates and are now thought to be too high.
\+ UN predictions (ICP, 1984).

In what may be regarded as a very obvious statement, but one which is often overlooked, the population growth rate at a point in time will have an impact upon provision of resources (food, housing, employment, education, recreation) for between 50-80 years into the future. Apart from the increased demand for resources created by a growing population further demand will be created due to the constant rise in affluence, living standards and expectations.

Planning for a 2.0% per annum growth across the entire human infrastructure would be difficult for even the wealthiest nations of the world. For the still-developing nations, such a growth rate is an impossibility (Brandt, 1980).

Even the most cursory study of population figures will show that the more we delay in bringing our global population figure into some sort of balance with the support base of the biosphere the sooner we face the prospect of a major crisis. The human population total is increasing according to the laws of geometric progression (exponential growth) whereas the resources of the biosphere are finite. The further we progress along the exponential population curve then so the greater become the demands upon the biosphere. Also, the time available in which we can restore a semblance of balance becomes shorter and more significantly, the reduction in time declines according to the rule of negative geometric diminution.

Nuclear Waste Problems

Modern man has an insatiable thirst for energy with which to power our urban, industrial and agricultural systems. Commoner (1969) calculated the energy consumption levels for each American to be in the order of 15,000 Kilowatt-hours per year whereas the biological minimum energy consumption per person is only 850 Kilowatt-hours per year.

Traditional sources of energy are finite (coal, oil, natural gas) and in the 1960s energy planners turned towards the nuclear industry as a long-term supplier of electrical energy. Early intentions were for 200 nuclear powered electricity stations to be operational in Western Europe by 1985 (Myers, 1985). In reality, this figure was only 73. Approximately 8% of the world's energy now comes from nuclear power although this figure hides massive variations. France, with 65% of energy from nuclear stations is top of the table; (Sweden and Finland, 40%; UK, 19%; US, 13%). Other nations such as Austria and New Zealand have declared the generation of power from nuclear stations to be totally unacceptable. New Zealand has adopted a total nuclear-free policy from the 1960s. Austria has constructed one nuclear power station at Zwentendorf (60km for Vienna). Its commissioning was rejected by the Austrian people in a referendum in 1978 and since that time the plant has been 'moth-balled'. The upsurge in 'Green Politics', see Chapter 9, combined with the nuclear power station disaster at Chernobyl in the USSR in

May, 1986 has meant that the Austrian Council of Ministers has agreed to the dismantling of Zwentendorf power station and the disposal of the unused fuel rods.

The nuclear industry is no longer regarded as being able to supply an acceptable form of energy. This is due to:

1. A series of nuclear accidents, ranging from minor leaks to two major disasters (Harrisburg, USA, 1979; Chernobyl, USSR, 1986).

2. Substantially under-estimated development, construction and operational costs have made nuclear generated electricity less competitive with traditional energy sources.

3. Falling energy demands in the light of the oil crises of the 1970s. Improvements in energy efficiency in oil consumption alone have led to 14% less oil being required in 1985 compared with 1975.

4. Problems of nuclear waste disposal.

Sweden has opted (1985) to dismantle its nuclear power industry by the year 2000 while in Britain, the Labour Party accepted a proposal at the 1986 Party Conference to remove all nuclear power stations by a similar date.

It is the long term problem associated with nuclear waste disposal (and nuclear accidents) which has caused a major rethink on behalf of governments and energy planners. There is still no acceptable solution to the problems of nuclear waste disposal. The spent fuel rods can be reprocessed to recover uranium and plutonium but highly radioactive wastes still remain. These wastes have so far been stored in protected tanks surrounded by water, buried in deep underground mines, contained in lead, concrete or glass shields, buried at sea or simply contained in concrete bunkers awaiting some future process to make safe the dangerous waste.

The ecological consequences of storing radioactive materials in concentrated 'dumps' pose major problems. Radioactive material represents an extremely concentrated form of energy with a very slow decay rate. Exposure to low concentration radioactivity for short duration (8 hour exposure) can lead to genetic and metabolic malfunction in biological materials (human cells) at some point in the future. The effects of exposure to radiation may not occur for between 5 and 20 years in the future; loss of hair, blindness, organ malfunction, tumours, cancers and sterility have all been attributed to radiation damage (Bookchin, 1974). Low level radiation may also be capable of changing the course of evolution through its impact on genetic recombinations.

All radioactive materials have 'half-life' values which indicate the amount of time necessary for radioactivity levels to subside to non-dangerous values. Some half-life values are very short. Iodine 131 has an 8-day half-life and after 50 days its radioactivity is but 10% of its fresh value. Plutonium 239, a product from all nuclear reactors, has a half life of 240,000 years and retains a lethal radioactive level 500,000 years into the future. Uranium 238 has a half life of $4.5x10^6$ years.

Faced with the practical problems of storing radioactive wastes for immense periods of time some energy planners are now reassessing the decision to 'go nuclear'. Of even greater significance has been the newly established strength of the ecological arguments associated with the problems of storing radioactive waste. Accidental leakage of waste and terrorist attacks on nuclear power stations are the two main reasons put forward against the widespread use of nuclear generated energy (Curtis and Hogan, 1980). An associated reason is the appreciation of the immense time factor associated with nuclear waste disposal. For the first time, society is questioning the wisdom of a contemporary action on the grounds that we have no reason to believe that a solution to the nuclear waste disposal problem will be discovered in the near future. This is just one small sign that man is beginning to appreciate the need to include the dimension of time in his problem-solving strategy. It may also indicate that our moral attitude towards the way in which we are prepared to treat our biosphere is gradually changing.

THE DEVELOPMENT OF AN ECOLOGICAL AWARENESS

Modern man, comfortably housed and adequately fed, is apt to forget that he is still an animal tied to the same basic requirements as an ant, a frog, or a cow. We must feed, breathe, excrete and reproduce in order that our species can survive. We have been able to achieve these life functions with considerable ease, hence the proliferation of our species. But far from becoming less dependent upon the biosphere and its resources, we are now more reliant on the biosphere. A larger population with greater expectations for its general standard of living has resulted in the use of more resources and the production of more pollutants.

An awareness of this problem gradually emerged in the 1970s. This resulted in research followed by recommendations on ways in which the biosphere could by protected from the most damaging actions of man's industrialised society. Green (1985) has observed that the newly found awareness of resource management and the need for conservation represents "a revolution in thinking about conservation that is perhaps as great as that in the war [Second World War] which led to our present structures". If this observation is correct then its implication upon our attitudes towards the biosphere and its components will be far-reaching.

A World Conservation Strategy (WCS)

Part of the revolution in awareness has been due to a major new initiative made by the combined actions of conservation bodies such as UNESCO, FAO, WWF, IUCN and UNEP. These groups were responsible for the publication of the *World Conservation Strategy*, (IUCN, 1980) in which a single integrated approach to global problems was presented to world governments. The *Strategy* was based upon three main propositions:

1. Such is the impact of mankind upon the biosphere that species and populations (both plants and animals) cannot successfully compete with man and must, therefore, be helped to retain their capabilities for self-regeneration.

2. A basic requirement for self-regeneration is the retention of the physical resources of the biosphere (climate, soils, water) in as un-

polluted state as can be possibly achieved within the limits of modern scientific knowledge.

3. Concern was expressed that the disappearance of many individual species of plants and animals was leading to:

a) a simplified trophic structure and

b) extinction of species would result in a serious loss of genetic diversity.

This would result in a possible simplification of future evolutionary diversification of species. This trend was contrary to what was deemed necessary if maximum flexibility was to be retained in our organic resource base.

One of the main recommendations of the WCS was that all future developments involving agriculture, forestry, soils, watershed management, extraction of minerals, sands, gravels, and native animal populations should be constructed on the basis of eco-development. This term is used to summarise two interrelated arguments:

1. Conservation strategies must be dynamic (as was shown on pages 8-11). If this was not so then the conservation policy degenerates into protectionism or preservation. Dynamic conservation implies sustainable development as a result of sound management policies.

2. Development of the human infrastructure has been based upon exploitive principles. There are indications that as a result of this strategy, 'shortages' of prime resources may occur in the future. (Serious local and short-term shortages have already occurred eg North Sea fish stocks, grain harvest in USSR, soil erosion in many north African countries.) Future development of the human ecosystem must incorporate a conservation strategy (based upon increased re-use of materials with level of usage set to an optimal level and a matching of natural renewal rate with the resource in question).

Eco-development

The concept of eco-development is not new. Its origins can be traced to the mid-west states of the

USA during the 1930s and '40s. Leopold (1949) first developed his concept of land-ethic early in the 1940s. He argued that in ecological terms operation of a land ethic involved a limitation on freedom of action in man's struggle for existence. In a philosophical sense, the land ethic is a differentiation of social, from anti-social behaviour. At the time Leopold developed his ideas on the land-ethic there was no formalised, acceptable structure dealing with man's relation to the land and its fauna and flora. The land-relation was essentially an economic relationship which involved concepts only of profit and loss, or as Leopold described it "entailing privileges but not obligations". The development of a land-ethic relationship was based upon the premiss that man, as an individual, was a member of a community of interdependent parts. As a strategic opportunist species our <u>instincts</u> are to compete and expand our position in that community but our <u>ethics</u> should prompt us to cooperate with our fellow species in order to optimise our chances of survival. Cooperation has, until quite recently, been confined to intra-species cooperation, that is, cooperation within our species in order to gain dominance over other species. The land-ethic as conceptualised by Leopold involves an extention of cooperation to an inter-specific level and to include all other animals, plants, water and soil, or collectively, the total living area, the biosphere.

It would be a fallacy to believe that the land-ethic concept is non-utilitarian in the sense that conservationists should argue on largely moral and emotive grounds that man has no right to destroy plant and animal species along with their habitats. More recent and more honest attitudes towards conservation and the land-ethic are anthropocentric and utilitarian in that they are involved with the maintenance of resources which ensure the well-being of man (Green, 1985). It is the acceptance of this latter approach which has given the conservation argument its greater acceptance by the non-conservationist members of society.

Eco-development has gained further respectability from the emergence of methodology which can evaluate the merits of different resource-use strategies. These techniques can be conveniently grouped into two, ecosystem (or landscape) evaluation (EE) and environmental impact assessment (EIA).

ECOSYSTEM EVALUATION

In order that eco-development policies may be applied to the biosphere it is necessary that the 'value' of individual components of the biosphere be measured so that a rank order of the components be established. Immediately, one is faced with deciding upon the unit of measurement to be employed. Table 4.2 gives the main evaluative approaches which can be used on biospheric resources. On their own, each method will provide a somewhat singular result and it would be advisable to select up to four evaluative methods and to combine the results in a comprehensive ecosystem evaluation survey. In this way the inevitable subjectivity which is contained in the individual methods in Table 4.2 can be reduced.

Pressure both upon the biosphere as a whole and on individual parcels of land can often be intense. This is particularly true in the old industrialised nations such as Belgium, Holland and Britain (Vink, 1975). Land space is at a premium and competition for land from a number of potential users is very high. Traditionally, the western capitalist solution to landuse competition is that the potential user who is prepared to pay most gains use of the land. This approach is based on the economic principle that whoever can afford to pay the most for a commodity will be able to achieve the greatest profit from that resource. Such a concept may be applicable to an industrial process but will inevitably be unsuited for application to a biosphere resource. Consider an area of European deciduous, hardwood forest. As an economic resource the hardwood timber comprising oak, beech, ash and elm would form a saleable commodity. A commercial forestry company would undoubtedly make a strong bid if such an area of forested land appeared on the market. The price it would be prepared to pay would be arrived at in the following way:

a. Calculate total area of woodland.

b. For a sample area calculate the number of different trees of commercial value.

c. Calculate the timber yield class of these trees and through use of a scaling factor, apply to total area of woodland.

d. Calculate total saleable value of timber.

e. Calculate profit:
(timber value - (extraction costs + purchase price of land)) = profit.

The balance sheet should not stop at point e. The deforested land could be replanted with high yielding exotic conifers, it could be converted to agricultural use and resold at a profit or it could be subdivided into housing lots and sold to a building developer again realising a profit.

Consider the same area of forest but from an eco-development approach. The value of the resource would now be assessed in a very different way:

a. Calculate the total area of the woodland.

b. Undertake an ecological survey in order to identify species composition, note occurrence of rare or threatened species.

c. Undertake an ecological evaluation, a yield evaluation, an amenity evaluation and a land use classification survey (Table 4.2).

d. Prepare a strategic use plan for the woodland so that a continuous supply of saleable timber can be produced along with a tree replanting scheme to ensure continuity of animal life and ground flora.

e. Establish a combined management/visitor centre and encourage school visits and scientific research.

f. A small number of holiday homes could be built on the edge of the forest, the revenue from which would help finance the conservation work.

The example of eco-development given above was designed to show that, with correct management, an ecosystem can provide a constant and varied output. Such a management plan would take many years to implement and would demand constant revision to cope with the natural dynamics of the ecosystem. If successful, the combined value of the different outputs, measured over a time span of, perhaps 50

TABLE 4.2 Ecosystem Evaluators.

Method	Example
Land Use Classification	Assessment of landscape units into one of a number of use categories (six or seven categories are usually sufficient). Categories defined by physical parameters e.g. soil, depth, elevation, slope (Kleingebiel et.al 1961, Bibby et.al 1977).
Financial Evaluation of Species	Allocation of notional monetary values based on abundance, conspicuousness and material values. Individual 'scores' are multiplied to give a 'shadow' price for each species (Helliwell, 1973).
Biological Yield Value	Assessment of volume of plant or animal protoplasm in an ecosystem. Involves complex measurement of biomass, productivity and/or production values (Jones, 1979).
Scarcity Value	The scarcity value of plant and animal species can be assessed using 'Red Data Book' methodology. Ecosystem scarcity assessed via a frequency assessment of ecosystem and/or landscape types (Perring and Farrell, 1977; Goldsmith, 1974; Walker 1970).
Ecological Value	Land units assessed on the occurrence of ecosystem habitat and/or community type. Involves 'expert' knowledge to determine units of highest ecological value. Geographic Information Systems can aid the identification of key sites (Davidson et.al, 1986).
Amenity and/or Scenic Evaluation	Landscape allocated a value based upon an aesthetic assessment. This can be a controversial method due to its subjectivity.

years, would be much greater than a once-for-all removal of timber, and sale-of-land strategy typified by the land speculator approach.

Unfortunately, man's working life span is too short at about 40 years for him to be concerned with long-term outputs. The 'quick-buck' philosophy is all too common in many facets of our modern life style and this is particularly so in the treatment we give to the biosphere. Human society understands the concept of 'scarcity values' when applied towards individual components of the biosphere (e.g. land ownership, yield values on fish stocks, forest, agricultural crops and animals). We have not yet been able to accept that individual component scarcity is a warning sign which indicates that we are approaching a situation where over-exploitation of the biosphere becomes a reality.

The misinterpretation of the 'scarcity signal' is analogous to the person who drives along a highway at high speed with a dash-board warning light glowing brightly. Such short-sightedness would inevitably lead to an expensive repair bill. Mismanagement of the biosphere will produce exactly the same result - a massive repair bill!

Detailed examples of the evaluation of plants and animals is given in Chapter 6. Concern with specific species whose future survival is in doubt is of very great concern but of even greater significance is the loss of habitats which support countless numbers of plants and animals.

As an example of the role that ecosystem evaluation can play in assessing the 'value' of a habitat we can examine the real-world problem of expansion of open cast mining towards an area of recreational land use.

Case Study: Blantyre Moor.

Blantyre Moor at first sight is an inconspicuous moorland located near the new town of East Kilbride in western Scotland (Grid Ref: NS 664 528). The moor is some 60 hectares in extent and lies 200m above sea level. Of its total area, approximately 23 ha has been partially drained to allow peat removal, see Fig 4.2.

Along the western edge of the moor there lies the small, but well used Calderglen Country Park while to the north east an area of open cast mining for coal and fire clay has gradually extended towards the moor. The nature of the problem as originally perceived was that the mining development could

Fig 4.2 Blantyre Moor, Western Scotland

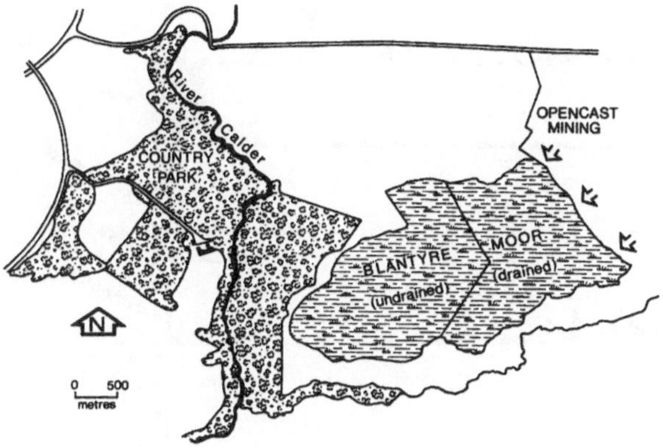

engulf the moor and finally threaten the character of the Country Park. If the moorland could be preserved as an 'ecological buffer' then the park would remain unaltered.

In 1984 an ecosystem evaluation of the moor was started. The following assessments were made:

1. Land Use Classification.
2. Scarcity value of site and species.
3. Ecological survey (floristic and habitat).
4. Pollen analysis of a peat core.

Assessments 1 - 3 suggested the moor was typical of many moorland areas in western Scotland (Burnett, 1964). Taken on their own the evaluative indices did not suggest the site to be of special significance. However, assessment 4, the analysis of fossil pollen showed the site to be of considerable scientific interest.

The moor was shown to be very deep, 10.75m maximum. Depth borings revealed the site to be a Pleistocene kettle hole with a perfect late and post-glacial sequence of fluvio-glacial clays and peats. A complete pollen analysis of the peat core was made, the results of which are displayed in diagram form in Figure 4.3.

Fig 4.3 Pollen Diagram for Blantyre Moor, Scotland

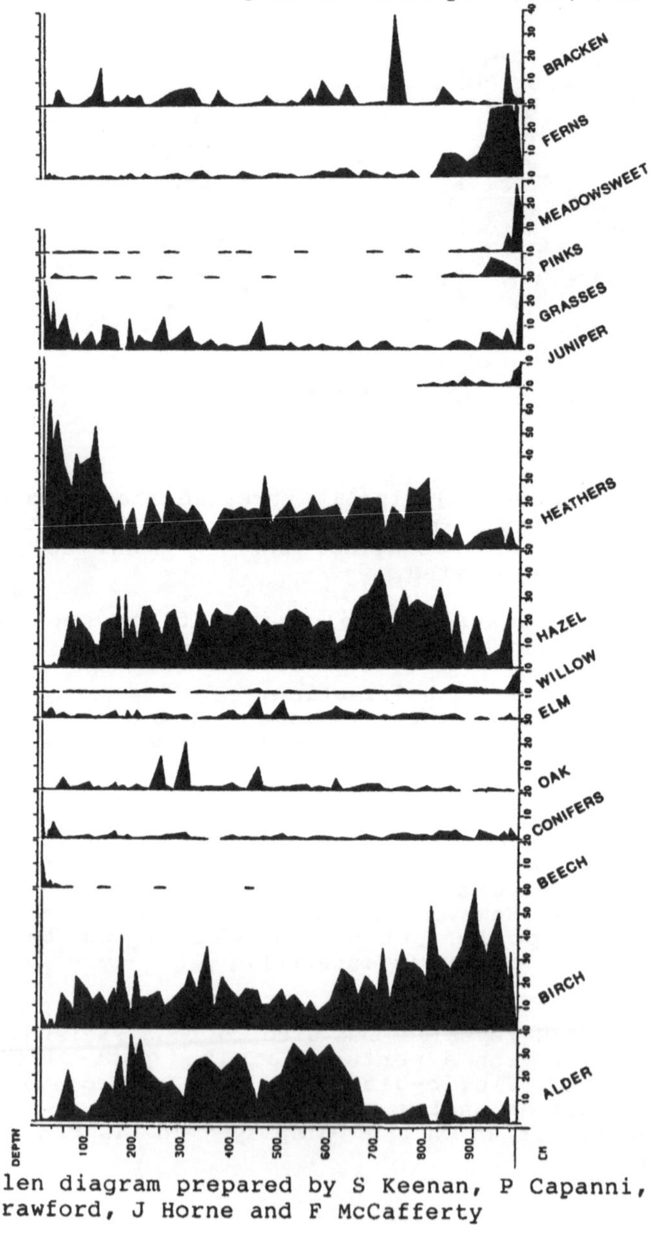

Pollen diagram prepared by S Keenan, P Capanni, L Crawford, J Horne and F McCafferty

Blantyre Moor was shown to have a clear sequence of post-glacial environmental and vegetation change. It is the deepest west of Scotland site to have been investigated in such detail and, as such, has assumed an important scientific value.

When judged on a combination of all four assessments, Blantyre Moor becomes worthy of conservation in its own right. It has been proposed for inclusion in the revised list of Sites of Special Scientific Interest.

The case study clearly illustrates the procedure necessary to establish the conservation-worthiness of an area. The process is not easy. It requires time, effort, research, enthusiasm and above all, cooperation from the existing landowner.

Ecosystem evaluation allows a site to be assessed on its basic criteria. It should not be a clinical diagnosis but should also include the opinions of a range of involved human beings including the land owner, land users and scientific advisers. By taking a responsible and balanced approach to conservation requirements, considerable progress can be made.

ENVIRONMENTAL IMPACT ASSESSMENT (EIA)

Ahmad and Sammy (1985) have claimed that no concise definition of Environmental Impact Assessment exists. EIA is a collection of aims, objectives and techniques which together allow the effects of a proposed anthropogenic action on the biosphere to be evaluated <u>before</u> the action has formally taken place.

In Chapter One the technique of ecological scenario construction was explained (page 20) and the need to exercise great caution in the interpretation of the scenario was stressed. EIA allows a much greater degree of precision to be applied to the forecasting of environmental and ecological issues.

An analogy can help explain EIA. Airline pilots are now trained on 'simulators' which mimic the specific behaviour of aircraft types. Take off and landing procedures at specific airports along with emergency drill can be practised in the 'safe' environment of the simulator chamber. Even though the simulator chamber may cost $10 million the advantages it gives in safer air transport, coupled with a saving in time necessary for pilot training is ample justification for the high investment in aircraft simulators.

How can the simulator concept be applied to ecosystems and their species? Imagine an upland terrain covered by mature woodland. The timber represents a valuable resource and a plan is prepared for its complete removal. What effect will a total logging policy have upon the soils, the drainage system, the associated forest vegetation and the animal population?

It would be easy to develop a subjective scenario in which deforestation was accompanied by soil erosion, gullying, changes in water quality run-off due to increased sediment load, disappearance of flora and the migration of animal species. Such a scenario would be basically correct in its main points, but in detail it would be hopelessly inadequate.

Forest ecosystems are the most spectacular and the most complex of all vegetation ecosystems on this planet. To construct a simulation model for a living, stable forest has been shown to be exceedingly difficult, (Duvigneaud _et al_, 1971). Borman and Likens (1971) in their now classic Hubbard Brook Experimental Catchment site were the pioneers in investigating ecosystem changes _after_ the change had been instigated. What an achievement it would be if we could accurately predict ecosystem changes _prior_ to a change in land use. If such methodology existed it would be possible to investigate alternative strategies in order to find the least disruptive change.

Natural systems have shown themselves to be very complex and most ecological simulations have so far been confined to the simpler agricultural system. Thus, Spedding (1975) produced a simulation study for grasslands and de Wit _et al_ (1971) a similar study for wheatlands. These simulation projects represented important 'learning stages' for ecologists and have led to far more ambitious simulation modelling of complex, probabilistic systems, for example, those described in Iyengar (1984).

Basic Concepts of E.I.A.

Munn (1979) has claimed that an EIA is:

> an activity designed to identify and predict the impact on the biogeophysical environments and on man's health and well-being of legislative proposals, policies, programmes, projects and operational procedures, and to interpret and communicate information about the impacts.

Because an EIA usually takes 'environment' to include all aspects of the natural and human environment it can validly study the flora and fauna, the climate, soils, the provision of employment, the state of general human health and the interaction between physical, biological, social and economic factors. Needless to say, an EIA of anything other than a very small problem, confined to a small geographical area soon becomes a very major undertaking.

In order that an EIA can make a recommendation it must be able to compare at least two different strategies. Ideally, the EIA would compare alternative strategies and possibly draw an optimum solution from several different strategies (Thirlwall, 1978). The optimum solution usually involves a compromise between what might be considered the ideal solution from a conservationist approach and the preferred outcome from the economist's or politician's view-point.

Compromise solutions often satisfy no one but if the proposed conclusion is shown to be based on sound premiss and is the result of a logical series of questions and answers then both the 'developer' and the 'conservationist' can usually be satisfied.

However, it has already been explained (page 101) that EIA is based upon predictions. It may be possible to use hard evidence from similar, existing situations elsewhere. For example, the proposal to site a conventional thermal electricity generation station on a tidal estuary would have statistical evidence from countless hundreds of pre-existing power stations located on estuaries. If, however, the proposal was to site an electrical generating station based on wave power or tidal power then the pre-existing data base would be confined to one or two working examples. Obviously, an EIA for the latter example would be able to offer far less conclusive information than for the thermal power station example.

Because of the great variety of human induced changes on the biosphere it would be an advantage if a common evaluative technique was available so that direct comparisons between alternative strategies were available. We should resist the temptation of making financial comparisons between alternative strategies (see Table 4.2) as ecological and conservation issues cannot be assessed solely in monetary terms.

The final point which applies to EIA is that its prime objective is to provide decision-makers (politicians, planners, local and national governing bodies) with information which allows them to arbitrate between conflicting land use proposals.

Conservationists and environmentalists may not readily concede the point but the decision-makers in our societies are rarely found in the 'environmental lobby'. If EIA is able to provide the environmental input which can be set against the commonly upheld economic input then this is an important step in the direction of accepting the environmental/conservation argument.

Development of EIA

Spellerberg (1981) has claimed that EIA is still in its infancy. As a technique it owes much to the development of computers as suitable means of storing, retrieving and manipulating the large quantities of data which are inevitable even in small EIA studies.

During the 1950s and '60s industrial, commercial and agricultural developments consumed increasing amounts of hitherto natural habitats. Species and habitat extinction proceeded apace (see Chapters 6 & 7). It was fortunate that local, national and international bodies recognised that such a trend was potentially harmful to the long term stability of the biosphere (Allen, 1980).

Just as the USA had been the instigator of the national park (page 115) then so too can that country be credited as the first to pass legislation requiring environmental impact assessment on major projects. The National Environmental Policy Act (NEPA) of 1969 became law on 1 January 1970. (See Chapter 8 for a detailed account of NEPA as a conservation planning milestone.)

Sandbach (1980) has shown that many of the early EIA projects were over simplistic. NEPA legislation required all major projects to include a clear environmental impact statement (EIS). To satisfy this requirement substantial numbers of ecologists and planners were required. One year after the enactment of NEPA, the Atomic Energy Commission had increased its staff involved with the preparation of EIS from 20 to 200. In 1975 the cost of producing all EISs stood at $162 million (Stoel and Scherr, 1978).

Many of the USA environmental impact assessments have become unduly lengthy and time-consuming to prepare. The Council on Environmental Quality (1975) admitted that many EIA's were too long, contained too much description and too little analysis. In spite of these criticisms EIA has been widely adopted by many other countries. Canada, Australia and New Zealand have enacted legislation very similar to the pioneer NEPA. Even when a formal requirement for EIA does not yet exist as part of the planning process (as in Britain) some international companies working in environmentally sensitive areas undertake their own EIA surveys. This is often done to convince the planning agency and the public of the concern which the company has towards the environment.

EIA Methods

Every project to which EIA is to be applied will demand an individual approach. A wide variety of methodologies exist; Ahmad and Sammy (1985) have claimed 86 distinct methods were distinguishable by 1976 in the USA alone. Clark et al (1978) have provided a thorough review of the different methods.

An early impact assessment was described by Nicholson (1971) for the Isles of Scilly in which a matrix was drawn across the main island and 'scores' for relevant data entered on the matrix.This method falls into the 'Leopold Matrix' group of techniques, see Leopold et al (1971) for full description. A complete Leopold Matrix comprises an open-cell grid of 100 possible existing environmental conditions on the horizontal axis and 88 environmental characteristics on the vertical axis. In most projects only a small number of the total combinations (8800) will be encountered.

Each interaction between horizontal and vertical axes is evaluated on a scale from 1 to 10. Weightings are used to indicate the relative importance and strength of interactions. The Leopold Matrix approach provides an exhaustive analysis to a problem but the final results cannot be easily or conveniently aggregated and their implications cannot easily be explained to a non-technical audience.

An alternative approach uses the geographical overlay technique often called the McHarg overlay after one of the first users of this method (McHarg, 1969). The area to be studied is overlain by a sequence of grid squares. Data for specific variables is then recorded for each grid square intersection or alternatively for the centre of each grid cell. For example overlays could be constructed for elevation, rainfall, soil type, industrial premises, distance to medical facilities. Superimposition of selected overlays would allow the user to identify distinct 'regions' in which specific combinations of variables occurred.

The overlay technique has the advantage of using a well known and easily operated method. Its main disadvantage is that when more than four or five variables are combined, the identification of regions becomes exceedingly complex.

In recent years the widespread availability of computers has revitalised the overlay technique.

Case Study. EIA Overlay Technique

Davidson and Jones (1986) have developed a Geographic Information System (GIS) for a test site of $70Km^2$ on the Campsie Fells near Glasgow in central Scotland. A data base was constructed for the study area in the following manner:

1. A 100 x 100 metre grid was superimposed over the area.

2. Data was collected for either the lower left

hand grid intersection of each grid square or a representative value for the whole square was calculated.

3. A total of 23 different overlays were constructed. Table 4.3 lists the range of data collected.

Table 4.3 Data Sets Collected for EIA. Campsie Fell Survey (data from Davidson & Jones, 1985)

1.	Grid Reference	13.	Trackways (P/A)
2.	Elevation	14.	Buildings (P/A)
3.	Slope	15.	Solid Geology
4.	Aspect	16.	Drift Geology
5.	Streams	17.	Soils*
6.	Rocks (P/A)	18.	Land Use Capability
7.	Marsh (P/A)	19.	L.U.C. Sub-class
8.	Roads	20.	Land Type
9.	Vegetation	21.	Exposure Value
10.	Tree Type		
12.	Reservoir/Land Outside Study Area		

(P/A) Data recorded on a Present/Absent basis

L.U.C. Land Use Capability

* Soil data formed a separate subset and was sorted on a micro-computer data base package

The grid square method of data collection is called the 'raster' method and it combines efficiency of data collection with computational simplicity. The main disadvantage of the raster method is that if a large grid square size is adopted - for example $1km^2$, then the data value for each variable becomes a crude approximation of reality.

A total of 230,000 data values were stored on a VAX mainframe computer. Short FORTRAN programs were then written to extract the required data and to produce symbol maps, Fig 4.4a. Initial maps were of simple, single variable distributions. Next, variables were combined so that composite maps could be produced, Fig 4.4b. Finally, interpretative maps were constructed, Fig 4.4c, in which algorithms were included in the program which searched the data base for specific variable combinations. Thus, prime agricultural land comprising those grid squares in which elevation was less than 100m, slope less than 5° and soil stoniness moderate to slight was plotted. Changing variable values was a quick and simple process and revised maps could be produced within an hour of working time.

Fig 4.4 Use of Geographic Information System to Establish Areas of Land Most Suitable for Conservation

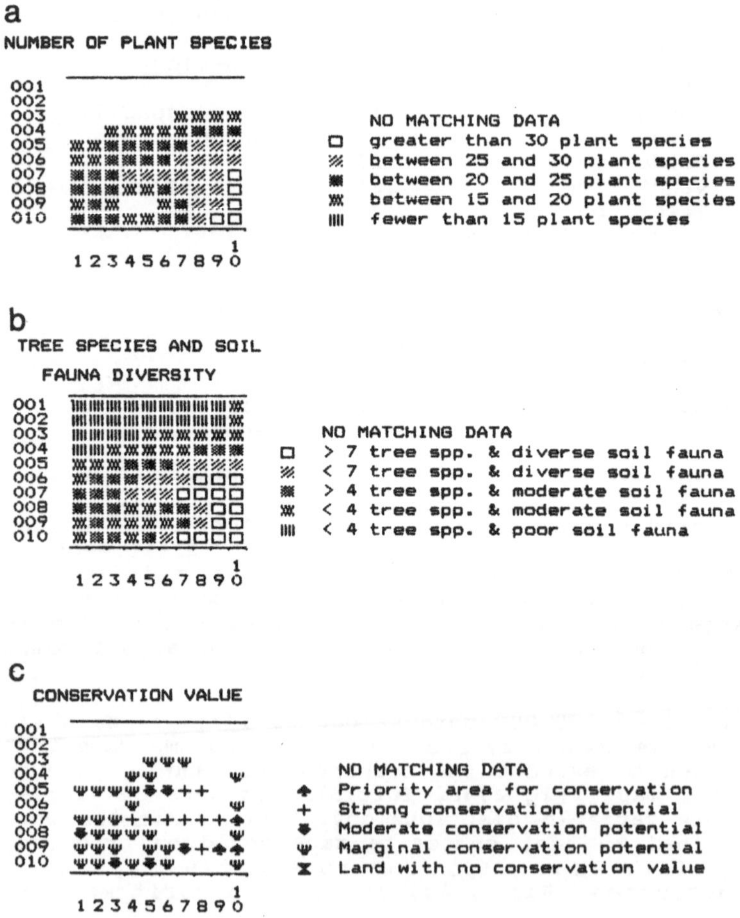

The advantages of this system were:

1. Specific maps could be drawn for variable combinations which the land use planner, farmer or forester considered of relevance.

2. The data base could be assembled from a wide variety of sources - field collection, air photos, existing maps, laboratory analysis, statistics, yield tables.

3. Maps could be produced quickly and relatively cheaply.

Disadvantages of the original system were:

1. The data base itself was slow to construct and consumptive of man-power.

2. The raster system gives an unreal or geometric display of information.

3. The original system was not 'user friendly' and was confined to a specialised mainframe computer.

The system, now renamed as MICRO-MAPPER has been transferred to a micro-computer system and is an interactive, user-friendly system.

The ability to construct interpretative maps to the requirements of the end user represents a totally new concept and has many advantages for the land planner and land user. It offers a flexibility of approach and permits the repeated testing of alternative strategies in land use. The technique has great potential for both developing countries where new landuses are being introduced and for old industrialised countries in which land use must follow the rules of maximisation due to high land values.

The use of EIA for forward land use planning should ensure that environmental considerations play a much greater part in determining the eventual use of land. The inclusion of economic and social data would allow the costing of alternative environmental and conservation strategies.

Chapter Five

CONSERVATION OF LANDSCAPES

Conservation of the biosphere and its species is now recognised by the majority of governments throughout the world as a fundamental necessity if the long term future mankind is to be secured. A dynamic conservation policy can help achieve this aim through the following:

1. The planned use of economic resources, notably the fossil fuels, but also extending to the use of minerals, plant material harvested directly from the biosphere, the waters and the air of the planet.

2. The retention of the maximum number of different plant and animal species in order to retain the genetic resource base of the biosphere.

3. Provision of recreation facilities in landscapes largely unaltered by recent advances in man's technology.

In order to achieve conservation of landscapes a wide variety of political, legal, social and scientific strategies have been devised. They have been applied with varying success by governments throughout the world.

Establishment of conservation status for landscapes is a phenomenon of the last one hundred years of man's history. The first area of land to receive this accolade was Yellowstone National Park in the USA in March 1872, (see pp. 114). Only in the last 25 years has international action been taken on conservation issues but even now, many governments still regard conservation as a concept of local concern as opposed to international significance.

Conservation Management

The need to conserve our ecosystems and species has arisen through a combination of mis-use and over-use of the biosphere by man both in times past and at the present. If mankind were to be suddenly removed from this planet, the biosphere would quite rapidly return to a more stable condition than that which currently exists.

Such a scenario is unlikely to occur and instead, deliberate conservation action by man has been proposed so that the damage caused both by our forefathers and by our own actions can be redressed. Unfortunately, we do not appear to be able to pursue a voluntary conservation strategy. Throughout the world it has been necessary for governments to legislate for conservation action.

International conservation agencies such as the International Union for the Conservation of Nature (IUCN) and the World Wildlife Fund (WWF), have done much to stimulate public awareness of the need for conservation. Equally they have been responsible for goading national governments into action. As a general target they recommend that each nation should set aside 10% of its land <u>and</u> sea territory for the exclusive use of nature conservation.

If conservation legislation is a pre-requisite for a conservation strategy, then conservation management is necessary to ensure species and habitat survival. Very few unaltered habitats exist in our biosphere. Simmons (1974) has shown that the proportion of 'unused land' (virgin land, or land which has been totally unaltered by man) has been reduced to a very small figure - probably less than 1% of the total land surface of this planet.

Land can be unused for two very different reasons. First, land areas may still be unexplored and uncolonised by man. These areas are now virtually non-existent; only the most inaccessible areas of Antarctica, or of rain forest regions are not periodically invaded by man. There are no new lands awaiting discovery. Satellite surveillance has allowed every hectare of land to be mapped in terms of its location, its land use, elevation and also its land resource potential. At best, this category of land can be termed 'under-used' or 'awaiting imminent use'. It can be land of the highest potential for conservation use as its species diversity is often very great.

The second type of unused land is represented by land of low resource potential. This comprises a large category of sub-types. For example, land may be unusable because of physical restrictions. Extremes of wetness, dryness, heat or cold can make many areas difficult for man's survival. Soil differences such as excessive stoniness, high salinity, low pH and vulnerability to erosion can make yet other areas unsuitable for use. Steeply sloping ground, high elevation and very exposed sites are yet other physical factors which inhibit use. Man may have despoiled other areas by previous use. Industrial and chemical wastes have poisoned much land, for example, the non-metalliferous wastes of the lower Swansea Valley in South Wales (Hilton, 1971). Our transportation links occupy many other areas of otherwise prime quality land, for example, motorway reservations, railway embankments and airports.

Wilderness Lands

Under-utilised lands have often been designated 'wilderness areas'. The concept of wilderness as landscapes which justify the effort of conservation has been best developed in the North American context (Nash, 1967; Tivy, *et al*,1981). In the USA the Wilderness Act (1964) enabled the setting up of the National Wilderness Preservation System. Additions to the 54 original wilderness regions have taken the total wilderness area in the USA to more than 50,000km. As the name suggests, wilderness areas are those areas of land in which formal land use is not recognised and no economic use of the land is permitted with the exception of presidential over-ruling of the Wilderness Act in times of national emergency. Minimum size of a Wilderness Area is 207km^2 (80 miles^2).

The enthusiasm for wilderness areas in the USA was evident by the passage of the *Wild and Scenic Rivers Act* (1968) which set out to conserve eight rivers 'with their immediate environments, which possess outstandingly remarkable scenic, recreational, geologic, fish and wildlife, historic, cultural and other similar values' (*National Geographic*, 1977). Progress in river conservation since 1968 has been erratic. This is due to the conflict which can develop between the rival users of a river and its immediate catchment. Under the Wild and Scenic Rivers Act, rivers or sections of rivers are classified according to a three part scale:

1. Wild - unpolluted, undammed with primitive surroundings, accessible only by trails.

2. Scenic - undammed, accessible by road with shorelines, largely undeveloped.

3. Recreational - readily accessible, with some development and pre-existing dams allowed.

Both the Wilderness Act and Wild and Scenic Rivers Act of the USA are major legislative bench marks in the formation of a network of special types of conserved landscape resources. Unfortunately, few of the other nations of the world can afford to follow the example set by the USA. Shortage of suitable land is an obvious problem for many of the old, industrialised nations, while economic restriction prevents the formation of wilderness parks in some developing nations. New Zealand has followed the example set by the USA. New Zealand's attempts to retain 'wild and scenic' rivers has resulted in Government funding of a major survey of waterway potential, both for economic and conservation uses (Egar and Egar, 1979; Jones, 1981).

A Comparison of Wilderness and Conservation Areas

The concept of wilderness represents one extreme view of a very wide range of conservation possibilities. For many people, the luxury of wilderness lands is considered an indulgence affordable only by nations with land resources to spare. Increasingly, conservationists recognise that it will become ever more difficult to establish extensive wilderness type areas and instead, greater resort will be made of a lower order form of conservation. This may by typified by the concept of the 'National Park' examined in detail later in this chapter. If existing areas of National Parks are to survive and new parks to be established in the developing world, then the 'value' of National Parks must be accepted by everyone, not just by ecologists and conservation activist groups.

The need to establish new parks is greatest in the Third World. Here can be found the few remaining tracts of under-utilised land. These lands are usually still richly endowed with native ecosystems and species. Unfortunately, Third World countries are also experiencing massive population increases and the consequent pressure on agriculture to expand its production has meant that under-utilised lands are being rapidly taken over for agriculture.

A REVIEW OF SOME NATIONAL CONSERVATION STRATEGIES

UNITED STATES OF AMERICA

Allen *et al* (1966) stated that:

> the greatest single problem facing the average U.S. citizen is good, socially acceptable use for his leisure time.

America undoubtedly represents the country with the highest standard of living, the greatest personal levels of wealth, highest car ownership levels and hence greater mobility factor. The inhabitants of the USA are not perceived by non-Americans to have problems over leisure time use; American football, Disney Land Leisure Centres, Las Vegas-style casinos, bars, clubs and a host of other organised leisure time activities should more than cater for the American needs for entertainment.

Fortunately, for the conservation movement, citizens of the USA are also a nation of campers, hunters, hikers and fishermen all of whom demand a high quality out-door recreation environment. Just why the American citizen lays such great emphasis on the outdoor environment is hard to determine. It may be a legacy from the days of the pioneer settler when a determined self-sufficiency was essential if a family were to survive the rigours imposed by settling a vast continental landmass.

The concept of allocating land to a specific and protected use was first seen in the Indian reserves set up in the 1700-1800s. Later, in 1845, Hot Springs in Arkansas was withdrawn from general use and dedicated to the future use of all the people. As far as can be told, there was no legal protection given to the Hot Springs reserve; it was the wish of the municipality that Hot Springs should be seen as a reserve area in which development was precluded. Some 27 years were to elapse before Congress created the world's first National Park, Yellowstone in March 1872. Legal powers were obtained to set up the park as a "public park or pleasure ground for the benefit and enjoyment of the people", (Allen, op cit). Yellowstone cannot, however, be judged as the beginning of scientific or elitist conservation because it was designated for the use of the public as an area of recreation and conservation.

Instead, the creation of Yellowstone National Park provided a major stimulus for conservation both in North America and in most other developed nations throughout the world. The prime function of Yellowstone was that of recreation; the spectacular scenic value of the area was already attracting many visitors in the 1870s and this trend has continued to this day. Following the success of the first park came a succession of others most of which were based upon a spectacular scenic resource such as mountain range (Rocky Mountain National Park), river system (Grand Canyon National Park), volcanic features (Lassen Volcanic National Park) or a biological resource such as a major native forest or swamp (Redwoods National Park and Everglades National Park). All parks created prior to 1930 were the result of withdrawal of land from the public domain and its consequent dedication as a National Park. Post-1930, land purchase and private donations to the federal Government have allowed new parks to become established.

The National Parks system of the USA has undergone constant expansion along with a gradual refinement of its legal structure. In 1916 the National Park Service was inaugurated to manage the growing number of parks. The National Park Service was the responsibility of the Department of the Interior and included monuments, battlefields, national cemeteries and other sites of scientific interest. The total extent of its policies now extends to some 60,000km^2. The *raison d'être* of National Parks was also changed in 1916. Their function was now seen to be:

> to conserve the scenery and natural and historical objects and the wildlife there in and to provide for the enjoyment of the same in such a manner and by such means as will leave them unimpaired for the enjoyment

of future generations.
(quotation in Simmons, 1974).

Another overhaul of the role of National Parks in the USA was undertaken in 1958 when Congress established the Outdoor Recreation Resources Review Commission (ORRRC). The purpose of the Commission was to evaluate the need for outdoor recreation up to the year 2000AD and to make recommendations as to how the needs could best be met. The function of the ORRRC was not primarily concerned with conservation issues. In order to fulfil its obligations ORRRC undertook 27 research projects and the most significant results are given below (ORRRC, 1962):

1. Between 1958 (base year) and 2000 the American population will double, demand for outdoor recreation will triple and quality of lifestyle, affluence level and leisure time will all have made significant but unquantifiable increases.

2. Base year recreation pursuits were identified as very simple, easily satisfied requirements e.g. walking, sightseeing, boating, cycling, picnicking, hunting, fishing, outdoor sports.

3. Future recreation pursuits will involve more demanding activities which require specialist facilities, for example, ski-ing, power-boating, hang-gliding, orienteering, and motor-sports. Traditional outdoor pursuits will still exist, though relative importance of individual activities will change.

4. Water is a focal point of outdoor recreation. Its use will increase both in terms of numbers of users and diversity of use.

5. An imbalance will inevitably exist between the location of National Parks and the location of centres of population. The effectiveness of the National Park system may thus be reduced.

6. Three quarters of Americans live in urban areas. Even with a good network of highways a demand for local areas suitable for outdoor recreation pursuits will exist.The annual vacation may allow more distant visits to be made to National Parks.

7. The advent of the caravan (mobile home) and the motel will lead to a major increase in mobility. Lengthened leisure times and an improved awareness of the natural environment will produce an increase in participation levels in conserved areas.

8. There are many overlooked and hence under-used sites suitable for recreation and conservation close to home.

A word of explanation over differences in the use of the term 'recreational land use' is appropriate at this point. Although the English language is the means of verbal and written communication in North America and in many other countries, some subtle variations in word interpretations exist. This applies to 'recreational land use' and 'national parks'. For example, in Britain a recreational land use is usually associated with a formal, organised sport such as football, horse racing or cricket. Only in recent years has the term gradually been extended to informal recreation such as wind surfing, climbing, pot-holing which previously have been classified as 'hobbies'.

In the North American context the term 'recreation' is interpreted in a much wider sense and includes long distance back-pack trails, canoe trails, hunting and fishing expeditions as well as the more formalised visitor centres providing information on short trails, wildlife and flora reserves.

The findings of the ORRRC were incorporated into the Land and Water Conservation Fund Act (1964), to be effective for 25 years from 1 January 1965. This act made available funds to develop outdoor recreational facilities provided that the state in which development was planned could raise 50% of the costs themselves. It was considered that this requirement would help assure citizen participation in the provision of recreation facilities. The response to the act has been good, with New York, California, New Jersey and several of the mid-west States financing large bond issues. In California, bond issues of $150 millions have been raised for conservation. It would appear that a large proportion of the population of the United States are prepared to finance an interest in outdoor recreation and not to rely exclusively on private commercial development or on Government sponsored schemes (Allen et al, 1966).

Although National Parks in the USA are strongly orientated towards recreation they are also involved in conservation work of the highest quality. Watershed management, protection of native forest and associated wildlife resources are major priorities. There are undoubted conflicts in interest between the recreationist and conservationist activities, particularly when the latter become elitist in nature. The National Parks Service is required to maintain the unspoilt nature of the parks yet at the same time must allow free public access. Visitors' pressure is growing rapidly, particularly as the number of visitors using the parks has been increasing at 9-10% per annum since the early 1960s. This rapid rate of growth (an example of exponential growth) has meant that visitor numbers have doubled every 7 years since 1960. Everhart (1972) has summarised the success of the national park in the USA as one in which emphasis has changed from the early days when the problem was how to provide sufficient facilities to encourage people to visit the parks to a situation in which the character of the parks is being lost beneath a growing surge of visitors.

Local Conservation in the USA

Apart from the widely publicised National Parks there exist a vast number of smaller, local examples of conservation. Each state is subdivided into counties, and in turn the counties are further divided into municipalities. Unlike other developed counties, the municipalities are allowed to develop their own approach to conservation issues. Most limit this to an interest in land use planning with specific zones being set aside in which conservation issues take precedent over commercial development. Local, and often small-scale conservation projects rely entirely on the drive and enthusiasm of municipal planners and are little influenced by state and federal government policies.

Problem Areas for USA Conservation Strategy

In spite of the undoubted lead which the American national park network has given to the rest of the world, Strong (1972) has indicated a number of problem areas inherent in the National Parks Service. Prior to the ORRRC investigations (page

116), park managers had assumed that the management of parks could be achieved through the implementation of 'sound ecological principles'. This, in many instances, involved non-management in the sense that ecosystems were assumed to be able to regenerate without interference from man. Traditional ecosystem succession theory as developed by Clements (1916) and Tansley (1946) was responsible for suggesting that, if left alone, ecosystems would attain a stable, self-perpetuating stage called climax. Whittaker (1953) and later May (1973) were able to cast doubt upon the earlier theories.

Areas of conserved land in the USA were facing the very obvious problem that they were coming under the influence of increasing visitor numbers and that ecosystem stability was not possible under the changing pressures which the national park system of the 1970s would generate (Strong, 1972). The problem was exacerbated by the following points:

1. The 1916 National Park Service Act and its subsequent revisions made no provision for a research capability.

2. The lack of a comprehensive and coherent research strategy for the park system as a whole.

The national parks of the USA have become so successful that most park managers now face the question of how much use can specific ecosystems withstand without incurring permanent and irreparable damage.

Another key area of concern involves unfavourable circumstances which occur outside the boundaries of the national park and which may have detrimental consequences inside the park. An example of such a problem can be found in the Redwood National Park of California which has been affected by commercial logging operations upstream of the park. Clear felling has resulted in changes in the peak flow rates of river water with associated changes in sediment load, erosion and deposition taking place within the park boundaries. Elsewhere, the extensive use of chemical fertilisers, pesticides, insecticides and herbicides in association with agricultural landuse has created problems of pollution drift into the park areas. Answers to these, and many other problems are seen by Rowntree et al (1978) as fundamental to the long term success of the American National Park System.

CONSERVATION IN BRITAIN

In respect of its geographical area and population size, Britain is the antithesis of conditions which prevail within the USA.

Britain is one of the oldest industrialised countries and has one of the highest overall population densities - 415 persons per km^2 or 1076 per square mile (Fothergill et al, 1985). As a result of many hundreds of years during which agricultural land use progressively extended over all but the highest, steepest and wettest sites (about 75% of the landscape is devoted to agriculture, Coppock & Best (1962)), little by way of natural ecosystems remains. When urban and industrial landscapes are added to the agricultural figure then very little land remains as potential 'wildscape'. Transportation and afforestation account for a further 10%. Military use of land has increased substantially and modern methods of warfare require much larger training areas. A modern battalion needs 35 000ha compared with only 157ha during the Second World War (Countryside Commission News, 1986).

It is, perhaps, surprising that suitable sites for nature conservation can be maintained alongside the commercial land uses. The establishment of a national park system in a small, heavily used landscape such as Britain has followed a very different direction from that in North America. Instead of the large, wilderness-style national park, owned by the Federal Government, Britain (or more accurately, England and Wales, for Scotland has no national parks) has relied on parliamentary legislation to carve out ten areas to which the title National Park has been applied. To the purist, Britain has no National Parks. As Burrell (1973) has correctly stated, the land designated in Britain as National Parks remains largely in private ownership, management is de-centralised, the land cannot be classified as wilderness (it is mostly under agriculture land use often of an intensive nature), and the main planning control is limited to that of building control. Due to the small size of the country and the widespread alteration of the landscape by man, the classical North American national park model cannot be applied and other forms of protection are required (Eagles, 1984).

The average size of the British National Park is only 1360km^2, (see Table 5.1). For the most part they consist of 'cultural landscapes' which bear the obvious impact of man's action.

Reference to Table 5.1 also shows that British National Parks are relatively young. The necessary legislation under which National Parks could be designated occurred in 1949 with the passage of the National Parks and Access to the Countryside Act. It was just one of the far reaching changes which changed the face of Britain at the end of the Second World War. Of even greater relevance to the landscape was the Town and Country Planning Act, 1947 which applied a new system throughout England

Table 5.1 National Park Statistics, England & Wales (from Bell, 1975)

National Park	Date of formation	Area km^2
Peak District	17. 4.51	1404
Lake District	9. 5.51	2243
Snowdonia	18.10.51	2175
Dartmoor	30.10.51	945
Pembrokeshire Coast	29. 2.52	582
North York Moors	28.11.52	1432
Yorkshire Dales	12.10.54	1761
Exmoor	19.10.54	686
Northumberland	6. 4.56	1031
Brecon Beacons	17..4.57	1344
Total area		13605

National Park land as a % of total land area of England and Wales	9.0%
English National Park land as a % of total land area of England	7.3%
Welsh National Park land as a % of total land area of Wales	19.7%

and Wales and greatly strengthened the administration of what was considered permissible land use. A separate though broadly similar Act applied to Scotland. Unfortunately, the 1949 National Parks and Access to the Countryside Act was not extended to Scotland as that country was deemed to be under less intensive land use pressure and contained a far larger area of semi-natural landscape, much of which was judged to be owned and used by a conservation-conscious land owning gentry. The 1949 Act also established the Nature Conservancy whose task was to establish and manage a whole series of additional conservation areas. These were to be confined to sites of scientific value. They were smaller again than the National Parks and differed also in that they were owned on behalf of the British Government by the Nature Conservancy, see pages 126-130.

The reasons for Britain's relatively tardy approach to national parks as a facet of landscape conservation has been accredited by Simmons (1968) as being due in part to government indecision and partly to the lack of any land in Britain which does not bear the obvious imprint of man. The paradox exists that while landscapes exist in Scotland which could justify the term 'national park' in an internationally accepted sense, Scotland has no national parks. England and Wales, which comprise landscapes made up of predominantly agricultural scenery have ten national parks, though it is relevant to note that the IUCN does not recognise these parks in their list of world national parks, see Table 5.2.

It is perhaps more remarkable that areas which bear the title of national parks exist at all in England and Wales. With intensive, commercial land use systems occupying the majority of land below 300 metres it is remarkable that landowners could be persuaded to give consent for part of their land to become national park territory. Pressure groups played a significant role, for example some long-established bodies such as the Royal Society for the Protection of Birds (1891) and the National Trust (1895) had long campaigned for protected areas of landscapes. Newer bodies such as the Councils for the Preservation of Rural England, Scotland and Wales were set up in 1926,'27 and '28 respectively and did much to focus attention on the changes taking place in rural Britain.

Table 5.2 Differences Between IUCN Definition of National Parks and British Definition from Phillips (1985)

IUCN definition of National Parks:

1. extensive natural areas
2. protected from exploitation
3. protected from occupation
4. the responsibility of national government
5. normally owned by the state

British (Anglo-Welsh) 'National Parks' are:

1. outstanding man-modified landscapes
2. in productive use
3. inhabited
4. the responsibility of local government
5. owned mainly by private individuals

In 1932 The Town and County Planning Act enabled, but not obliged, local authorities to plan the use of the countryside (Bell, 1975). The act was a negative one in that it could prevent undesirable events from taking place in the countryside but it was very limited in its ability to establish conserved areas. Public pressure groups formed the Standing Committee on National Parks in 1935 and a Scottish equivalent was formed in 1943. Government interest was slight until the advent of the Second World War caused a major rethink in the ways in which land use could be optimised. Despite the locking-up of resources and man-power in the war-effort, time was found to operate a whole series of far-looking committees whose task it was to plan for post-war reconstruction. The Scott Committee (1942) on Land Utilisation in Rural Areas made it clear that the establishment of nature reserves and national parks would be of benefit to a whole range of potential users within Great Britain.

The Dower Report (1945) provided the foundation upon which The National Parks and Access to the Countryside Act (1949) was based. Dower's definition of a National Park was as follows:

An extensive area of beautiful and relatively

wild country in which, for the nation's benefit and by appropriate national decision and action

a) the character of the landscape is strictly preserved

b) access facilities for public open-air enjoyment are amply provided

c) wildlife and buildings and places of architectural and historic interest are suitably protected, and

d) established farming use is effectively maintained.
(quoted from Bell, 1975).

This definition bears a marked similarity to that of the 1916 definition which was prepared for the occasion of the creation of the National Park Service in USA (see pages 115). Dower's definition includes an additional point in which farming interests are to be protected within park boundaries. This point was essential for two reasons:

1) to pacify the fears of the farming lobby that their agricultural practices within National Parks would be forcibly altered

2) to give protection to farmers from large numbers of urban dwellers insensitive to the problems of the farming community who would 'invade' the national parks at weekends and during holiday periods.

One further committee, the Hobhouse Committee (1947) reported prior to the eventual passing of the National Parks Act. Hobhouse confirmed the recommendation made by Dower. Twelve national parks extending for some 14,716km^2 were proposed along with 52 'conservation areas'. The latter were defined as 'areas of high scientific value with considerable scenic value'. The fact that separate conservation areas were to be designated implies that the function of the national parks was not primarily that of the safeguarding ecosystems and species of ecological (scientific) value. Equally, the conservation area (which underwent a change of name to Area of Outstanding Natural Beauty), was not intended for use as a public recreation area.

However, due to their scenic attractions, combined with a location at or near coastal locations, (for example the Gower Peninsula in South Wales), AONB's have been subjected to some of the most intensive human use.

Ten of the twelve proposed parks were established, see Fig 5.1 and Table 5.1. The two areas not deemed appropriate for national park status were The South Downs and the East Anglian Broads. In the case of The Broads, reclamation costs were considered to be too expensive and for The South Downs, degradation processes due to intensive agriculture was considered to have proceeded too far. Two other areas have been subsequently considered for national park status, The Cornish Coast (1950s) and Cambrian Mountains, Mid Wales (1972). The Cornish location comprised small, discontinuous areas with high administrative costs, while the Mid Wales proposal was rejected because of the large number of objections from existing landowners.

National Park status in England and Wales bestowed administrative powers, firstly to the National Parks Commission (1949-1968) and post-1968 to the Countryside Commission. The Commission has the responsibility for

> reviewing all matters relating to the provision and improvement of facilities for the enjoyment of the countryside of England and Wales, the conservation and enhancement of its natural beauty and amenity, and the need to secure public access for open air recreation.
> (Report of the Countryside Commission, 1968).

The Sandford Review (1974) confirmed many of the earlier considerations. It recommended setting up a co-ordinated park management structure in order that a more realistic approach to park management could be achieved. No changes in the political or planning infrastructure as related to national parks and conserved areas was considered necessary.

Other Forms of Conservation in Great Britain

National Parks in Britain are cultural landscapes with an above average score for quantity and quality of 'natural areas' (Eagles, 1984), and within which public, open-air enjoyment can be maximised. These

Fig 5.1 Proposed (1947) National Parks and Conservation Areas
(after Ministry of Town & Country Planning 1947)

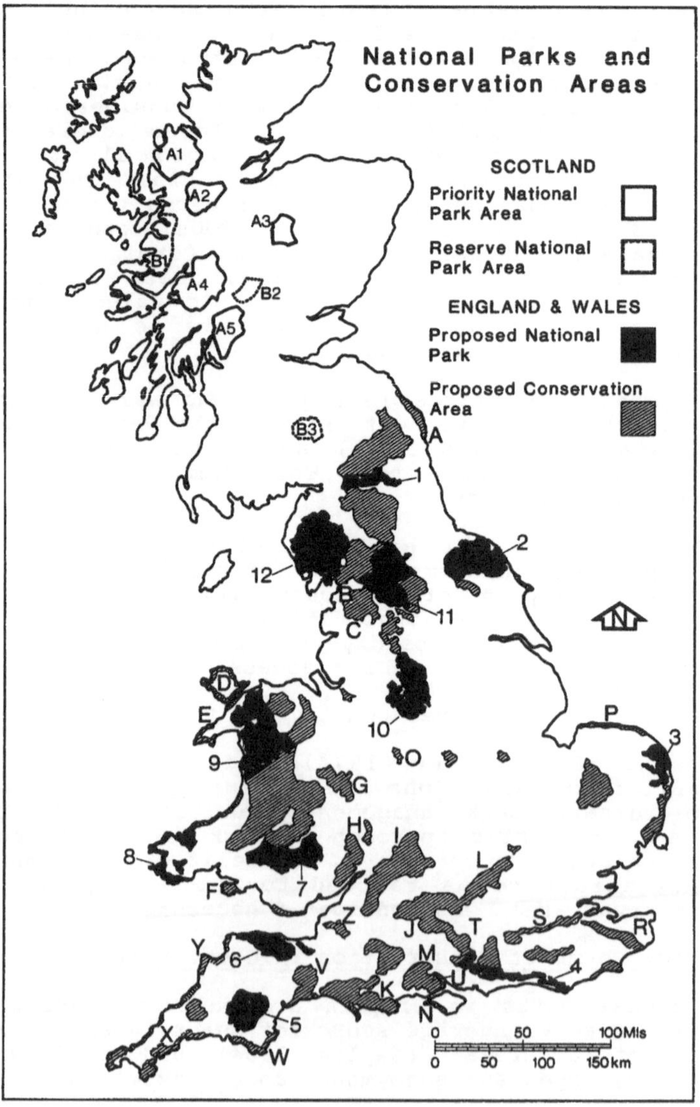

Fig 5.1 cont. Legend

England & Wales - Proposed National Parks

1. Roman Wall (Northumberland National Park)
2. North York Moors National Park
3. Norfolk Broads National Park (disgarded)
4. South (Sussex) Downs National Park (disgarded)
5. Dartmoor National Park
6. Exmoor National Park
7. Brecon Beacons and Black Mountains National Park
8. Pembrokeshire Coast National Park
9. Snowdonia National Park
10. Peak District National Park
11. Yorkshire Dales National Park
12. Lake District National Park

England & Wales
Proposed Conservation Areas (renamed AONBs)

A. Northumberland Coast
B. Arnside & Silverdale
C. Forest of Bowland
D. Anglesey
E. Lleyn
F. Gower
G. Shropshire Hills
H. Wye Valley
I. Cotswolds
J. North Wessex
K. Dorset
L. Chilterns
M. South Hants Coast
N. Isle of Wight
O Cannock Chase
P. North Norfolk Coast
Q. Suffolk Coast & Heath
R. Kent Downs
S. Surrey Hills
T. East Hampshire
U. Chichester Harbour
V. East Devon
W. South Devon
X. Cornwall
Y. North Devon
Z. Mendip Hills

Scotland - Priority National Park Areas

A1. Lochs Torridon & Maree, Little Loch Broom
A2. Glen Affric, Glen Cannich & Strath Farrar
A3. Cairngorms
A4. Ben Nevis, Glencoe & Black Mount
A5. Loch Lomond & Trossachs

Scotland - Reserve National Park Areas

B1. Moidart, Morar & Knoydart
B2. Ben Lawers, Glen Lyon & Schiehallion
B3. St. Mary's Loch

areas, important as they may be in retaining the character of the landscape, do not satisfy the more demanding requirements put forward on behalf of scientific-based conservation.

It is fortunate that at the same time as legislation enabling national parks to be set up in England and Wales, action was also taken to establish a network of some 150 National Nature Reserves (NNRs) and 3500 Sites of Special Scientific Interest (SSSIs). The strategy behind the setting up of these sites has been explained in two documents, Conservation of Nature in England and Wales, Command 7122 (Ministry of Town and Country Planning, 1947) and National Parks and the Conservation of Nature in Scotland, Command 7235 (Ministry of Town and Country Planning, 1947). Nature conservation in Britain would only be assured through the establishment of a number of key sites selected to represent

> all major types of natural and semi-natural vegetation, with their characteristic assemblages of plants and animals and habitat conditions of climate, topography, rock and soils and biotic influences. (Ratcliffe, 1977).

Different from the US National Park Service Act (see pages 115-119) in which no provision was made for a research capability, Command 7122 gave specific emphasis to the requirement of research into the careful management of the scientific and nature conservation values of the chosen sites. Furthermore, the research was seen to further the advancement of scientific knowledge of the conserved ecosystems and this knowledge would be utilized in the management of the twelve (later reduced to ten) national parks.

Command 7122, (op cit), was remarkably far-sighted in its recommendations and conclusions. It recognised that because of the increasing influence of man in Britain no ecosystem, nor plant or animal species could be assumed to be safe from destructive forces. The same document also stressed the need to make 'reserves' as large as was practicably possible. Figure 5.1 shows the geographical extent of the proposed conserved areas made in Command 7122. Unfortunately, many of these sites were omitted from the 1949 Act which set up National Parks in Britain.

Ratcliffe (1977) has explained the procedure whereby sites worthy of conservation could be identified. The overall strategy was to establish a series of sites which provided an acceptable representation of all the natural and semi-natural ecosystems in Britain. Seven main ecosystem groups were identified:

Coastlands	Lowland Grasslands,
Woodlands	Upland Grasslands and Heaths,
Peatlands	Heaths and Scrubs
Open Waters.	

For each group further selection was undertaken on the following criteria:

1. Size (extent) - the concept of viable size was of major concern. The conserved areas must be neither too small (fragmented), to allow the ecosystem to survive, nor must it be so large as to make management too difficult.

2. Diversity - the conserved area should possess a wide range of habitat conditions (ecological gradients) so that an equally wide range of species type and number can be supported within the conserved area.

3. Naturalness - no areas of truly natural habitat exist in Britain. This category might be better termed 'degree of modification'. Those areas which have been least modified by man are generally most worthy of conservation.

4. Rarity - of great relevance in conservation management. Rare habitats and communities are also often fragile and require a particular sensitive management strategy.

5. Fragility - a complex criterion but involves the sensitivity of a site to undergo change. Communities undergoing seral succession are highly vulnerable to change whereas communities which have reached an equalibrium point with their environment tend to be more stable.

6. Typicalness - an index of how well an ecosystem may be judged to represent its type.

7. Recorded history - sites at which scientific

study and research have been undertaken are sometimes of 'proven value' when compared with sites which have not been studied.

8. Position in ecological/geographical unit - it was considered desirable to include as many sites of ecological value within a given geographical area.

9. Potential value - some sites could develop, with appropriate management, a high nature conservation profile.

10. Intrinsic appeal - it was recognised that human interest in some ecosystem types was greater than for others.

Selection of Areas for Conservation

The identification of areas suitable for conservation was a formidable task. It was achieved through the work of regional groups of Nature Conservancy staff whose task it was to map the occurrence of every semi-natural site at the 1:25 000 scale. For each site an evaluation of conditions was necessary to allow the grading of the site into one of four categories of conservation, or a fifth group labelled unworthy of further consideration. Problem sites or disagreements between conservationists and land owners would be referred to a scientific assessor.

The four conservation grades are:

Grade I sites were of international or national importance, and would usually be the equivalent of National Nature Reserves (NNR). These sites should be afforded the maximum conservation status.

Grade II sites were judged to be only slightly inferior to Grade I sites but may duplicate some of the features to be found in Grade I. They can be regarded as 'alternatives' to Grade I sites.

Grade III, sites of high regional importance and usually designated Sites of Special Scientific Interest (SSSI).

Grade IV, sites of lower regional significance, still worthy of SSSI status.

It is important to stress that of the remaining areas of land and water in Britain, many areas although not being worthy of inclusion under the conservation categories listed above, would play a vital role in providing a total living space into which plants, and particularly animals, would stray for varying periods of their life history. A conservation strategy cannot be confined to legally protected reserves. In order that a conservation policy can work successfully, it demands that landowners and the general public act in a manner which is sympathetic to the needs of conservation. To this end, education and the political system play major roles.

Areas of Outstanding Natural Beauty (AONBs)

Yet another level of conservation can be found in Britain. The all important Command No.7122 (Ministry of Town and Country Planning, 1947) was clear that while the purpose of the NNR was primarily that of scientific conservation it was equally adamant that an attempt should be made to conserve landscapes of special 'amenity' value. It was proposed that Conservation Areas be created within which regions of 'distinctive character' be identified and managed in ways to preserve their unique character and heritage. As for National Parks, the Conservation Areas were intended to apply only in England and Wales. A change of name to Areas of Outstanding Natural Beauty occurred prior to the first designation being applied to the Gower Peninsula in May 1956 (Bridges, 1986). A total of 32 AONBs have been established, ranging in size from the smallest (Dedham Vale - 57km ^{2}) to the largest (North Wessex Downs - 1738km^2). The latter site is larger than most of the National Parks. AONBs have no strict legal status but designation has allowed some planning authorities to gain a tighter control over land use and building regulations (Bush, 1973).

Long Distance Footpaths

Britain is covered by a complex network of ancient foot-paths, tracks and roads many of which traverse the remotest areas of Britain. Many tracks link Celtic look-out and religious sites. The National Parks Act (1949) recommended the integration of some of these tracks into the National Park and Conservation Area network.

Particular difficulties have been experienced in the siting, maintenance and management of the footpaths. Opposition from land owners has also been a problem in some areas. Walkers are seen to require camping sites, leave litter, break fences, leave gates unsecured, bring dogs into proximity of livestock, cause erosion problems and disrupt the otherwise secure living areas of upland birds, mammals and plants.

The first and best known footpath was the Pennine Way, 402km long, and opened in 1965. Since that time new paths have been opened at a rate of approximately one per year (see Fig. 5.2 for distribution).

Conserved Land Use in Britain - Conclusion

The modern phase of conservation of landscapes and ecological resources in Britain started much later than in many countries. However, since 1947 an elaborate legal and scientific structure has been established to pursue the implementation of a conservation policy. There are, unfortunately, still areas of deficiency, the most obvious being the lack of National Parks in Scotland. Conserved areas in Britain are now under severe pressure from agricultural and forestry land-use changes and from visitor pressure. To this end, the Nature Conservancy Council and the Countryside Commission have begun (1986) a new conservation survey of Britain in which a computer-held data base of scientific and land-use information will be used to re-assess the need for a conservation policy.

Finally, no study of conservation in Britain would be complete without mention of the numerous trusts, societies and conservation groups who are responsible for a wide range of conservation work. At the largest scale the National Trust have been responsible for purchasing several key areas which have escaped the net of the formal conservation process. Thus, Cadaer Idris in Wales, and Ben Lawers and Ben Lomond in Scotland are all under the management of the National Trust.

Fig 5.2 Long Distance Footpaths, Great Britain

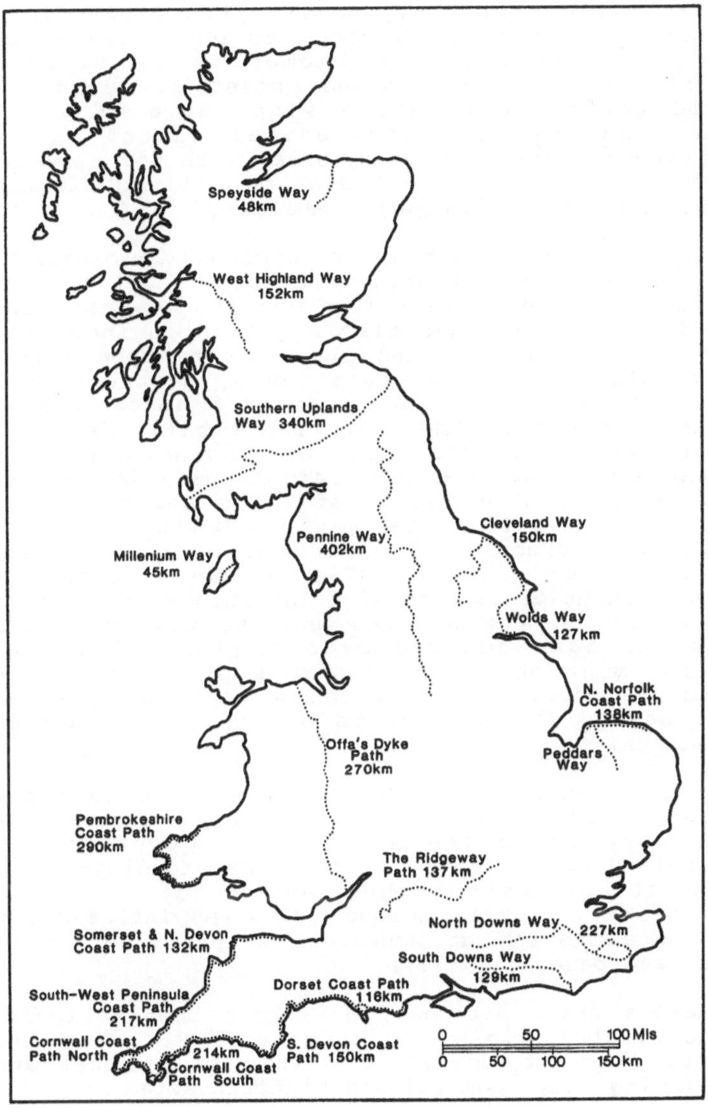

CONSERVATION IN SOUTH AMERICA

The continent of South America contains some of the last vestiges of truly natural ecosystems to be found on planet earth. It also possesses a huge range of terrestrial biomes, from tropical rainforests through lowland, moist grasslands, to arid environments and a wide range of alpine environments. The latitudinal extent of the continent (12°N - 55°S) combined with the elevation range provided by the Andean mountain chain have given an infinite range of habitats.

South American countries are exclusively classified as developing nations. Without exception they exhibit rapid population growth, catastrophic inflation rates, and also a growth in industrial development which places great demand upon primary raw materials. As a consequence, destruction of natural ecosystems has proceeded apace and innumerable ecosystems have been totally destroyed along with the total loss of unknown numbers of plant and animal species. This tragedy is made all the worse by the fact that this destruction is occurring within the Amazon Basin - an area generally acknowledged to be one of nature's most diverse areas (Bourne, 1978). It has traditionally been thought of as a natural laboratory in which the evolution of new species and the development of hybrids has occurred over the millennia. The despoilment of one of nature's prime regions of original development has consequences which extend far beyond the boundaries of the South American landmass.

The emergence of a modern conservation attitude began in 1903 when Lake Nahuel Huapi in the Andes was donated to the Argentinian government by Francisco Perito Moreno on the understanding that it was to be established as a national park (Costantino, 1972). Moreno had been influenced by the conservation attitudes which prevailed in the USA (see pages 114-119).

Numerous other national parks have been designated since that time but the term appears to have been misused as it covers a multitude of parklands including city, coastal and historic sites.

The influence of North American parks policy was further imparted following Latin American participation at the First World Conference on National Parks, held in Seattle during 1962. A Latin American Committee on National Parks was established in 1963 and fostered a series of action programmes. Since that time, progress has been slow and conservation and national park management still occupies a lowly level usually within Forest Service Departments, or occasionally under the control of the Director of Tourism (as in Costa Rica). Lack of finance is perceived to be a major restriction on conservation work.

The Peruvian response differs from other South American countries. Areas judged to be worthy of conservation are provided with legal status and are set apart from development plans. Overseas scientific aid is then obtained which enables an inventory of the area to be made. If the survey results are favourable the area may be opened to the public, or held under a 'protected status' until some future occasion when funds may be available for management work to be completed.

Of the South American nations the efforts of the Ecuadorian government to establish the Galapagos Islands as a centre of world heritage is to be commended. The Columbia Government also undertook a major review of the national conservation requirements during 1974-76 (Wetterberg & Meganck, 1978). The nation provides a good example of the large range of environmental types to be found with Latin America (see Table 5.3).

The Columbian conservation programme was started in 1967 and has as its objective the reserving of one representative sample from each major biome type found in the nation. By the late 1970s, 1.5 million hectares had been established as the Columbian National Park System (1.2% of the total land mass). The system comprises the following categories of land:

> National park - an area of size such that it can be ecologically self-regulating. Minimal disturbance by man. To be retained for scientific, educational, aesthetic and recreational values.
>
> Nature reserves - an area in pristine condition of flora, fauna and landscape. Dedicated to conservation, research and study of its natural riches.

Unique natural area - area of special or rare flora or landscape.

Faunal sanctuary - area dedicated to the preservation of animal species and their genetic resources.

Parkway - highways and adjacent strips of land which possess singular panoramas, natural or cultural values and is conserved for education or relaxation.

Table 5.3 Environmental Zones in Columbia. (after Wetterberg & Meganck, 1978)

Elevation in metres	Temp. range (°C)	Extent (Of land area)	Bio-climatic region
0-900	20-23	85%	hot
900-2100	23-17	9%	temperate
2100-3000	17-12	6%	cool
>3000	<12	2%	paramo

The Columbian National Park system has made rapid and successful progress during its short existence. It provides a good model for other South American countries to follow. Wetterberg (1974) has claimed that the achievements of the Columbian conservation movement are more credible when judged against the poor perception for the need of a conservation strategy by South American governments and public. The often rapid turnover in governments which hold extreme political views also leads to major discontinuities in conservation policies. Land ownership is jealously guarded by South American families for the status and security it bestows. The severe financial restrictions which prevail throughout most of the continent prevent expenditure on low-priority options such as conservation. The same author has suggested that the intense nationalistic pride to be found in South American countries could possibly be turned to the advantage of conservation if national governments could be persuaded of the 'status value' of natural ecosystems and their attendant wealth in terms of gene-banks, scientific research and foreign exchange potential via tourism.

CONSERVATION IN NEW ZEALAND.

The conservation policy of every nation will depend upon a host of historical and contemporary facts which serve either to dispose towards or against the general ethics for conservation of species and ecosystems. A case study of New Zealand illustrates this statement very clearly.

Human impact on the biota of New Zealand was late to arrive. The isolated, oceanic location bestowed a remoteness which had existed for many millennia. The flora and fauna of New Zealand has a high proportion of endemic species (Godley, 1975). It has been assumed by Salmon (1975) that its isolation from evolution in the rest of the world was due to its remote geographical position. Visits by Polynesian sailors were occurring by 1000AD (Davidson, 1976) and by 1200AD a considerable human population existed. The impact of the first settlers was similar to that of all human colonists; deforestation was followed by the establishment of grasslands for grazing herbivores. In New Zealand there were no herbivorous ungulates to fulfill this role, it was filled instead by herds of grazing birds - the remarkable flightless moa (Fig. 5.3) which reached formidable size (2m high). When the European settlers reached New Zealand, in the 1840s, the moa was already extinct, hunted by the Polynesians for its meat, feathers and bones.

The Polynesian Maoris had far less impact on ecosystems and species than most other human cultures because the Maori showed a sympathy and concern for the land, its plants and animals to an extent unseen in most other cultures. The European settler was, in contrast, land hungry. Persuaded to emigrate to New Zealand by the promise of good, cheap land, the white settler was soon to devastate the native forest and its animals and replace them, for the first time in the history of New Zealand, with grazing mammals. This destruction occurred at about the same time as conservation was taking its first steps in North America (Yellowstone National Park, 1882). In 1887 a Maori chief, Te Heuheu Tukino, concerned at the continual loss of Maori Fig

5.3 Two Bird Species Unique to New Zealand

a) The Moa (extinct by 1840s) <u>Dinornis maximus</u> height approx. 2 metres

b) The Takahe <u>Notoruis mantelli</u> height 250 mm

lands to the European farmers, donated the summits of three volcanic mountains in the North Islands to the Crown for 'the purpose of a National Park' (Lucas, 1972). It is not clear whether the North American concept of a national park had reached New Zealand by 1887, or whether the action by the Maori chief was of independent origin.

Legislation had been formalised by 1894 and New Zealand obtained its first National Park - Tongariro National Park. Four other parks followed in the period up to 1928. In 1952 a National Parks Act created a National Parks Authority as the body responsible for the "scientific, conservation and recreational functions" of the park system. Total park area has increased to 10 parks covering 2,050,922 ha. (some 8% of the land area), see Fig 5.4. The parks have been selected to include the widest range of natural ecosystems. The largest park, Fiordland National Park (1,223,347 ha), is located in the most southwesterly extremity of the South Island. This area contains a number of extremely rare birds including the takahe (<u>Notoruis</u>

Fig 5.4 National Parks of New Zealand
data from Cumberland & Whitelaw (1970)

Name	area ('000 ha)	Date of formation
1. Urewera	199	1952
2. Tongariro	67	1887
3. Egmont	33	1900
4. Abel Tasman	18	1942
5. Nelson Lakes	57	1952
6. Arthur's Pass	98	1929
7. Westland	84	1952
8. Mount Cook	70	1952
9. Mount Aspiring	199	1952
10. Fiordland	1223	1952
	2050 = 8% of NZ land area	

plus 2.59 million ha of scenic reserves, historic reserves and bird sanctuaries.

plus 25.9 million ha state forest and 16.8 million ha of Crown and Maori land.

mantelli, (see Fig 5.3b) a species thought to have been extinct until its rediscovery in 1949. An interesting and unusual feature of New Zealand national parks is the scientific study of introduced species on indigenous plants and animals, for example the effect of deer, wild pigs and possum. Some of the national parks have a limited tourist function; Fiordland National Park has the famous Milford Track which allows hikers to follow a way-marked trail for four days through the wilderness landscapes of the park (Patterson, 1978).

After a short-lasting destructive phase, the white settler in New Zealand has quickly adopted a strongly conservation-conscious life pattern and numerous action groups now pay close attention to the requirements of additional conservation needs throughout the land and water resources of the islands.

CONSERVATION IN AFRICA.

In the context of the African continent conservation is synonymous with the vast savanna game parks populated with large numbers of grazing ungulates (buck, zebra, giraffe) and made famous by the production of high-quality television documentary programmes. The real situation regarding conservation in Africa is very different. Most conserved areas are located in areas which are unsuited for development by both the European colonist and the indigenous population. Thus, areas which are very arid or are disease ridden (tsetse or malaria infested) have been designated as national parks or 'reserves' (Curry-Lindahl,1972).

African nature reserves and national parks do not represent natural ecosystems but show many features of man-made interference. For example, the Queen Elizabeth National Park in Uganda is regularly burnt and all efforts to prevent this have been unsuccessful (Owen, 1973). Recognising the deficiency of conserved areas in Africa, the African Convention for Conservation of Nature and Natural Resources was set up in 1968 specifically to safeguard 'those ecosystems which are most representative of, and particularly those which are in any respect peculiar to, their territories'. In addition some 350 plant and animal species were listed for special conservation treatment (Curry-Lindahl, 1972).

The most common African national reserve is the grassland and park savanna type biome. Many of these reserves are very large in size, for example the Kruger National Park in South Africa (20720 km^2)

and the Serengeti National Park in Tanzania (14500 km^2). Both these examples were originally established as game reserves, for example Kruger National Park had been the Sabie Game Reserve since 1898 and was upgraded to full national park status in 1926. Nowadays, the best known African national parks have become major foreign currency earners with international tourists being escorted in air-conditioned luxury between top class accommodation centres. In some African parks the visitors can be assured of seeing and photographing large mammals because game wardens arrange feeding points at strategically located points along the tourist route.

Table 5.4 Foreign Exchange Earnings from Tourists, Kenya (from K. Curry-Lindahl, 1972)

Year	Amount (millions of US Dollars)
1960	19.2
1964	25.2
1967	28.8
1968	44.0
1969	47.4
1970	53.2

There is considerable argument over the merits of basing the conservation of African wildlife upon the desires of rich tourists. In the elitist conservation sense tourism and conservation are anathema, they must not mix. In the hard light of reality, however, many of the African nations are desperately poor and if tourism can generate an income (see Table 5.4) and at the same time bring about the creation of national parks, then such a strategy becomes attractive. Some African nations operate a 'split-system' of conservation for tourists and conservation for scientific purposes. This is particularly so for Zaire, Kenya, Tanzania, Uganda, Zambia and South Africa. Under this system tourists are encouraged to visit some sites whereas other sites receive little publicity, public access is restricted and park accommodation is not available. Zaire has an exemplary record of scientific conservation and this has been maintained despite the political and economic turmoil which has beset the country. Similarly, South Africa has a sound policy of allowing tourists into approximately 10% of its total reserve area, the remaining 90% being for scientific conservation.

Conservation areas are most numerous in southern and central Africa with quite major deficiencies in the north and northwest of Africa. Nigeria had, until quite recently, few areas of conserved land. Apart from the Olokemeji Forest Reserve which was 'gifted' to the British monarch in 1900 there were no conserved areas until the Yankara Game Reserve was created in 1956. Nigeria's first national park was established in 1975 and is an amalgam of two pre-existing game reserves (Borgu and Zugurma) along with Kainji Lake which is man-made (Ojo, 1978). Management of the park for tourism and for ecosystem conservation does not appear to be well established.

National parks, game reserves and forest reserves throughout Africa are subjected to numerous pressures. These are varied and diverse in nature. Population increase has made great demands for additional agricultural lands; value of biotic resources has made poaching of animals for hides, meat and ivory a huge international industry (see page 165) while natural hazards such as drought, floods, overgrazing by herbivores, soil erosion and disease all undermine the work of conservationists in Africa.

Nowhere is the need for an active conservation policy based on sound ecological principles greater than in Africa. Only through the achievement of an ecological balance can many of the continent's problems be solved. The devastating droughts of the 1980s have brought additional difficulties for conservation in the form of species losses, soil erosion, fire damage and perhaps of even greater concern, have financially impoverished parts of the continent. Financial resources over the next decade will inevitably be spent upon the rebuilding of the agricultural ecosystem. Great care will be required to ensure the maintenance of a conservation ethic.

CONSERVATION IN THE USSR AND EASTERN EUROPE

The history of conservation in the Soviet bloc has a long and creditable record. Prior to the 1917 revolution, many nature reserves or 'zapoviednik' had been created. The establishment of new reserves began immediately after 1917 with 250,000ha on the shores of Lake Baikal being designated (Heptner, 1978). These newly conserved areas were established with the prime aims of protecting threatened species, particularly the mammals and birds. This policy was extended in 1930 to include some vegetation species - the Pitsunda pine and the yew. The average size of zapoviedniks was between 20-40,000 ha though some very large ones were also created, (e.g. Pechjora-Ilych 721,333 ha.). Even greater areas of taiga and tundra could have been designated but it was considered unnecessary because they were under no threat of modification. Reserves are most common in the south-east of the Soviet Union in the Manchurian-type mixed forest.

Later, collaborative work between USSR, Romania and Poland has resulted in many rare and threatened species showing numerical revivals, e.g. the European beaver.

Two interesting and very different features of Soviet conservation practice are the following:

1. The insistence on 'restricted access' to the reserves (Pryde, 1972). Tourism and recreation in the North American custom are not allowed. The reserves are for scientific use only. In recent years there has been a departure from this approach and on the shores of Lake Baikal an 'open' national park has been created.

2. Conservation in the USSR is treated as a 'hard-science' and there exists a plentiful supply of research information and a large and highly skilled group of conservation science personnel attached to research institutes and under the overall control of the Central Laboratory for Nature Conservation - Ministry of Agriculture.

The Soviet and Polish approaches to conservation of ecosystems and species are probably the best example of what can be termed the elitist approach to conservation. Few other nations have considered it possible to pursue this approach. While it appears that the elitist approach can achieve conservation results quickly and effectively it suffers several disadvantages:

1. It can be operated only where extensive 'empty' land spaces exist.

2. Because of its purist approach it requires a rigidly planned society in which to operate.

3. It is carried out in isolation from public involvement.

It is the latter point which makes the Soviet approach to conservation unacceptable to western nations. Public involvement through education, field visits, tourism and recreation are all integral parts of the western approach to conservation. It would appear that the somewhat rigid Soviet approach can be very successful and achieve the desired conservation result very quickly.

Chapter Six

THE CONSERVATION OF SPECIES

The Species and its Environment

Plants and animals do not exist in isolation. They live together in complex groupings called 'communities' in which species are inter-connected to varying degrees by competitive and cooperative bondings (see Table 2.7).

Very few species are solitary by nature. All animals must come in contact with others of its species for the purpose of reproduction (an intra-specific relationship, see page 45) and it will also meet other species whilst feeding, resting and nesting (inter-specific relationships, see page 44). Animals also come in contact with plants - again for purposes of feeding, nest building, shelter and for the intangible requirement of attainment of territory (see Fig 6.1).

Plants and animals also co-exist in complex relationships with the environment which surrounds them (see Chapter Two). Plants and animals must obtain all the necessary resources for the successful pursuit of their life history from the environment.

It is surprising that such a complex, interdependent set of relationships between biological, chemical and physical components can actually work in such harmony and without assistance from any obvious guidance system. The organism-environment relationship is the result of gradual, evolutionary development. Through a process of trial and error over the millennia, countless thousands of ecosystems have been produced which are ideally suited for the long-term survival of organism and habitat. The

Fig 6.1 Relationship of the Species to the Surrounding World

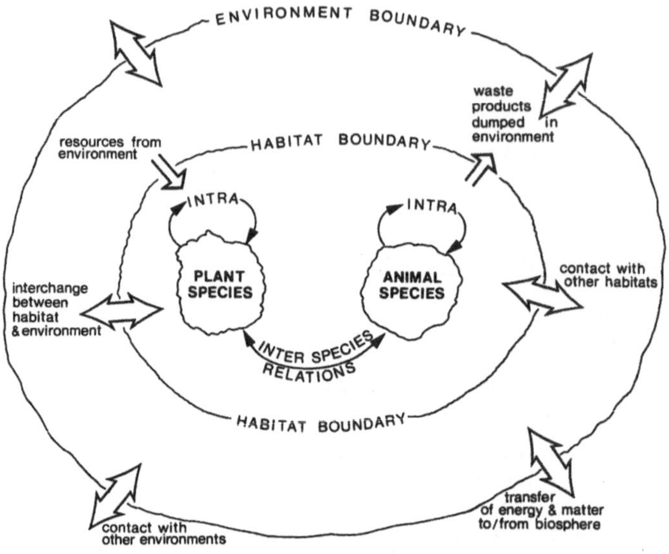

ecosystems are held in position, each relative to the other by means of an intricate control system in which feedback signals are constantly shaping the ecosystem for its optimum chances of survival in the prevailing conditions (see pages 47-52).

Environmental Threats to Species Survival

Our biosphere, comprising parts of the atmosphere, lithosphere and hydrosphere is undergoing constant change. On the shortest time scale are the diurnal changes bringing variation in light intensities and temperature patterns. Diurnal change is repetitive and predictable; it causes behavioural change to occur (for example, sleeping patterns) but little else.

Seasonal change, based on the movement of our planet about the sun causes additional repetitive and predictable change, but on a time scale measured in weeks or months. This type of change is effective in controlling the growing cycle and in determining migration behaviour in animals.

Seasonal change is responsible for some biotic catastrophes. The onset of the first frost of autumn will kill many of the summer annual plants. Many insects caught without their protective winter adaptations may also be killed. But because seasonal changes are repetitive and have occurred many times before, their effect on life forms is not catastrophic. The impact of the changes has been incorporated into the life histories of the species concerned.

Major biospheric alteration such as climatic change along with isostatic and eustatic changes in land and sea levels and even continental drift are also occurring. Can these macro-changes pose a serious challenge to the survival of species? All these changes occur so slowly that plants and animals can respond either by migrating to more favourable sites, or by adaptive changes which allow the species to retain its competitiveness within the new environment (Keeton, 1980).

Circumstances which are potentially damaging to the biota are those which occur suddenly and without warning. They are usually of limited geographical extent, for example, a land slip, a flood or a gale. Of somewhat greater spatial extent are volcanoes and earthquakes which can devastate many hundreds of square kilometers of land. The effect on the biosphere of the Mount St Helens volcano (Brown, 1986) was such that severe disturbance of the region occurred with substantial species mortality. As far as can be told, however, no species has been made extinct by the Mount St Helens volcano.

Natural catastrophes can bring about dramatic biospheric change but the magnitude of that change gradually diminishes and eventually stability is resumed. The effects of the catastrophe are healed with time and it often requires considerable effort to distinguish the previous ecosystem damage, (Fig 6.2).

Of the naturally occurring events, climatic change has probably the greatest impact upon the biosphere (White et al, 1984). Whenever climatic change is accompanied by continental drift then it is possible

Fig 6.2 Landslip Sequences Over Time, Arthur Valley, Fiordland National Park, N.Z.

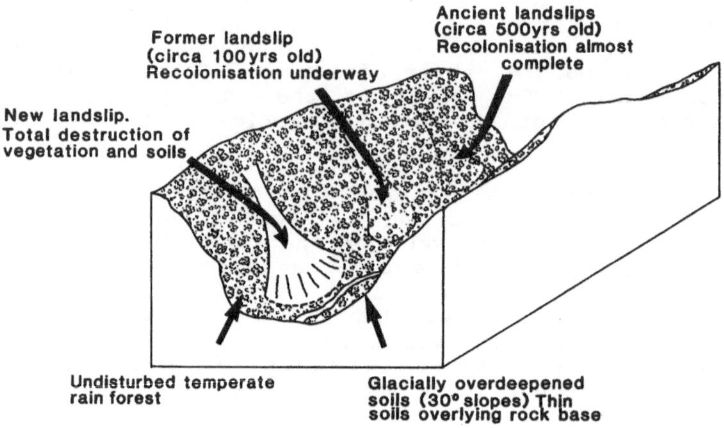

for a species to find its habitat undergoing rapid and quite radical change (Simmons, 1979). But even the relatively major changes are normally insufficient to exterminate species. Instead, the species is able to adapt to the changing conditions. Thus at some time in the past a single species of the genus now called Nothofagus (the Southern Beech) found its geographical distribution split into three distinct areas due to the effects of continental drift. Separate land masses drifted apart to form South America, Australia and S.E.Asia and New Zealand. Nothofagus can be found in all areas today, but in each area separate evolution has occurred in response to regional peculiarities.

A more spectacular adaptation in response to changing environmental circumstances can be found from the dinosaurs, a distinct form of reptile which dominated the animal kingdom during the Jurassic era which began 180 million years ago and lasted for about 45 million years (Wells, 1960).

Based on the fossil record of dinosaurs it appears that this group disappeared completely some 65 million years ago and thus would qualify as a natural extinction of species. However, recent work by palaeo-zoologists and palaeontologists suggests that dinosaurs did not become extinct but underwent a rapid phase of adaptation to suit them to new climatic, geological and biotic conditions which prevailed in post-Jurassic times. Certainly, the large dinosaur typified by Diplodocus, Triceratops and Tyrannosaurus species disappeared but totally new forms appeared which today have resulted in our avian fauna (birds) and a group including crocodiles, lizards, snakes, turtles and tortoises (British Museum, 1979), see Fig 6.3.

Fig 6.3 Cladodiagram Showing Possible Relationship Between Dinosaurs and Living Archosaurs from British Museum (1979)

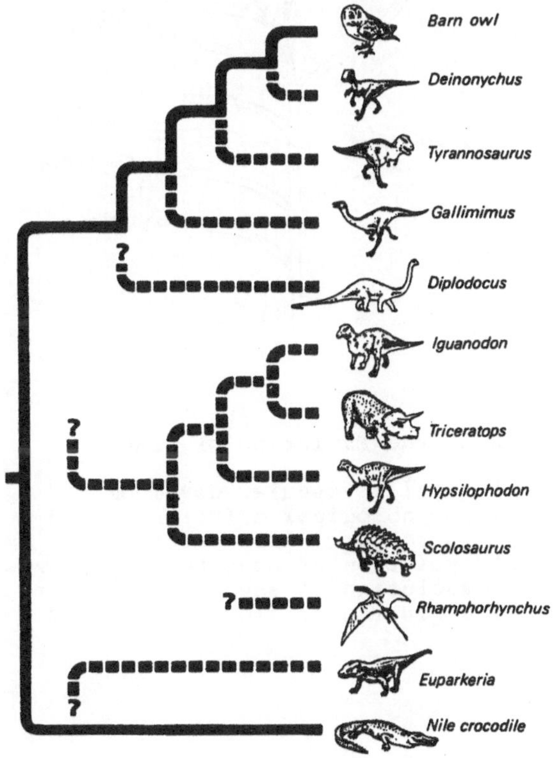

Plants too, show an equally distinct pattern of emergence, dominance and ultimate replacement by newer and more dynamic species which are better able to thrive in the prevailing conditions. But as was shown for animals, the superceded plant species are not entirely forced into extinction, (Fig 6.4). The most unsuccessful of the 'recent' plants appear to have been the Gymnosperms (the cone-bearing plants).

Fig 6.4 The Evolutionary Sequence of the Plant Kingdom (from Eyre, 1970)

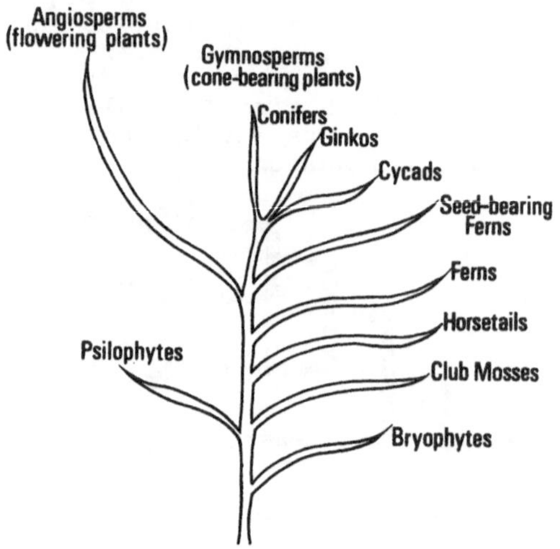

Stages in the Evolution of Plants

a) initial vascular development
b) morphological differentiation
   i) roots   ii) leaves
c) evolution of cone bearing plants
d) evolution of seed bearing ferns
e) evolution of flowering plants

An examination of the present flora and fauna on this planet leads to the conclusion that organic profusion and diversity is greater now than at any time in the past (World Resources Inst., 1986); see Table 6.1. Such variety holds great promise for the future development of our biosphere. Based upon our current understanding of biosphere productivity and diversity the biotic resource base appears to be still expanding.

Table 6.1 Number of Species by Class (from World Resources Inst., 1986)

Class	Identified species	Estimated species
Mammals	4170	4300
Birds	8715	9000
Reptiles	5115	6000
Amphibians	3125	3500
Fishes	21000	23000*
Invertebrates	1300000	4400000
Vascular Plants	250000	280000
Non-vascular Plants	150000	200000
Rounded totals	1742000	4926000

* minimum figure

Unfortunately, the continued expansion of biospheric diversity must be tempered by a comparatively recent trend which, if allowed to proceed unchecked, could lead to a reversal of biosphere expansion. Within the last 10,000 years one species has emerged as the overall dominant and by so doing has caused imbalance, extreme competition for space and resources along with the deliberate extermination of many species which are judged to be 'enemies'. The species responsible for this destruction is, of course, mankind (*Homo sapiens*).

Other species have in times past held dominance, for example the Trilobites in the Cambrian era and the Lycopods and tree ferns during the Carboniferous era. No other species apart from ourselves has been so successful in selectively hunting, persecuting and removing other members of the plant and animal kingdoms. No other species apart from mankind has been as successful in modifying the environment and habitat conditions and by so doing has thrown into jeopardy the survival of countless thousands of plant and animal species. Ehrenfeld (1970) estimated that extinction due essentially to man's actions in the 12,000 years of post-glacial history has exceeded the total number of extinctions brought about by the previous one million years of time during which our planet experienced a major ice age - the Pleistocene era.

Lucas and Synge (1978) have suggested that 25,000 plant species throughout the world may have been brought into the 'at serious risk' category through the inadvertent activities of man. That figure is no more than a 'guesstimate' based upon detailed work undertaken in small areas. It represents an extinction rate of 10% of the total plant species.

Usually, it is not the species themselves which are directly threatened but their habitats. Thus, the Mediterranean Basin, small islands, rain forests and deserts along with islands in the Indian Ocean are under greatest threat of destruction. On the Hawaiian Islands the highest threat value of all has been recorded (Lucas and Synge, _op cit_, where habitat destruction has placed almost 50% of plant species under threat of extinction.

The last of the planet's great natural vegetation regions, the Tropical Rain Forest, is now being exploited for its natural resources with scant regard for conservation of habitats, fauna and flora. Allen (1980) has suggested that by the year 2000 AD habitat destruction within the Tropics (22.5^{o} north and south of the equator) will be so severe that up to one million plant and animal species will disappear! That figure will comprise mainly the lower orders of plants and animals (the bryophytes, mosses, invertebrate animals and decomposer organisms) many of which have not yet been scientifically named nor their life histories studied. Many of these species may pass into extinction without us ever knowing their role within the biosphere.

Man the Problem Species

Homo sapiens would probably win first prize in the

competition to find the least ecologically conscious organism on this planet! Tivy and O'Hare (1981) have no illusions about our attitudes towards the biosphere having described our species as a "dirty animal". Indeed, there are times when we appear intent on creating situations which will eliminate all life forms on this planet. Nuclear warfare of even modest proportions (for example, a one-for-one exchange of air-borne nuclear warheads between two nations) would devastate life forms over much of an area as large as north-west Europe and would have serious implications for the stability of the entire biosphere. Perhaps our political leaders should rename our species. We are no longer behaving according to our Linnean name, *Homo sapiens*, which can be translated to 'group of wise beings'.

Not all extinctions are due to the direct and deliberate actions of man. The draining of swamps in lowland, central African countries has been undertaken to eradicate the breeding ground of disease-carrying insects such as malaria and the tsetse fly (Highton, 1974). The improvements to human health and to cattle have been immense but a corollary of this work has been the elimination of lowland swamp habitats and the extinction of countless aquatic species.

The inadvertent destruction of habitats under the guise of improving the chances of survival for mankind has been a constant feature of both recent and early history of man. Examples can be found from every continent. Today, the Bengal tiger in India and the Australian kangaroo are threatened with extinction in the same way as the wolf was threatened in the period 1400 - 1750 AD in Europe and the giant flightless bird of New Zealand, the moa (*Dinornis maximus*) was taken to extinction by the Polynesian settlers in the pre-1700 AD era.

The behavioural attitudes of *Homo sapiens* towards the well-being of the biosphere and its contents is one of continuous exploitation. The paradox is that our species has achieved its present status by adopting a selfish attitude to the rest of the biosphere. Without this attitude *Homo sapiens* would have remained merely another member of the animal kingdom.

Changing Attitudes

The motivations for a change in our attitudes towards the need for conservation have been considered in detail in Chapter One. They can be summarised under the general headings of economic, practical, ethical and political reasons (although not necessarily in that order).

Above all else it now seems probable that many of the species which are threatened with extinction may have a resource use for mankind. This use may be as a material resource, or used for food, in medicinal preparations (see Appendix A) or increasingly may be used in what is termed 'genetic engineering' (Myers, 1983) - the combining of genetic material in laboratory conditions to create unique species and varieties which are of particular value to mankind.

It has become accepted that before further major alteration of habitats and loss of species have occurred we should embark upon a major scientific assessment of the total value of a species both as natural components of ecosystems and as resources for man. This will not be an easy task and methodologies for evaluating the total value of species are still in the earliest stages of development.

Valuation of Biotic Resources

Human civilisation has, since the very earliest of times, used the biotic resources of the biosphere with scant regard for the future supply and security of those resources. Early European man hunted deer, the North American indian hunted buffalo, the aborigine the kangaroo. Forests have been cleared for agriculture and the replacement anthropogenic system has led to massive soil erosion. If we could in some way measure the value of the biosphere resources which have been used we would be able to recognise the size of our indebtedness to the natural resource base.

Unfortunately, no assessment of any cost has been made. The very real problem exists of how can we evaluate a herd of deer, a shoal of fish or a forest? Would we place a higher value on one of the last members of the North American bison than on one of its very numerous ancestors? In other words do we increase the price for 'rarity value' as an antique dealer would do for a rare item? Or should we assess the value of a species on the basis of its ability to breed and thus provide us with a renewable organic resource? If the last example were true then a fast breeding rabbit might be valued more highly than a slower breeding goat.

There is no simple and generally acceptable way of evaluating the flora and fauna, though as a later section of this chapter will show there have been several attempts to develop an objective evaluation scheme.

Any valuation of the biota will inevitably involve the calculation of its financial price. We are accustomed to accepting financial indicators as guides to the state of national economies. Thus, we have indices of monetary supply, inflation rate, bank rate, gross domestic product, share indices and a host of other economic evaluators with which to judge the health and wealth of national economies (World Resources Inst., 1986). It is quite natural that we should look for another quantifiable index with which to measure the state of the biosphere.

Unfortunately, no such index currently exists. We do not know which parameter we should use to indicate the value of the biosphere. Should it be productivity, rarity, quantity, calorific value, genetic purity or genetic adaptability? Or should we develop a subjective questionnaire with information about aesthetic value, ethical treatment of plants and animals or pleasure to be derived from looking at the biosphere in an unmodified form?

Many conservationists abhor the placing of financial values on a biotic resource, preferring instead to use an aesthetic or scientific value. Unfortunately, industrialists, economists and politicians cannot equate this type of non-financial valuation with a traditional fiscal balance sheet in which income and expenditure can be translated into neat columns of profit and loss.

It is possible that if such a financial balance sheet was prepared it might reveal a biotic asset of such wealth that the owner would prefer to realise the asset than to conserve it! This is indeed the situation for many of the developing nations of the Third World. Equatorial rain forest represents a valuable resource base which can earn essential foreign exchange from the industrialised, northern hemisphere countries (FAO, 1982). Is it naive to suggest that if the inhabitants of the wealthy industrialised nations want equatorial rain forest to remain then they must pay, not to fell the trees and convert the timber into valuable veneers for use in the furniture market, but instead to conserve an area of rain forest in its natural state?

There would be many difficulties to such a scheme. Could a government of a wealthy, developed nation be persuaded to pay a politically unstable country not to clear an area of rainforest? Would the temptation on behalf of the developing nation to exploit the situation lead to over-pricing, or even multi-charging for the same piece of forest? How would the partners in the agreement come to a mutually acceptable value for the resource and for how long should the contract last? Should the conservation be absolute or should some use of the resource be allowed by the indigenous population? Finally, how could the policy be adequately policed?

As yet, no bi-national or multi-national agreements of this nature exist. There are, however, very many national arrangements whereby a government will fund the purchase of land for conservation use. National parks would be a prime example whereby a political decision is made to support conservation. In the early 1980s the British Nature Conservancy Council entered into medium-term financial agreements with private landowners whereby land of known scientific value would not be brought into commercial farming. Peat and fen bogs have featured mainly in these agreements. Judged by the amount of critical correspondence that has appeared in the press it would appear that this type of arrangement does not yet meet with general public approval.

A development of the financial arrangement with the land owner can be seen in the designation of six areas in England and Wales as environmentally sensitive areas (ESAs). Incentive payments will be provided for farmers in the ESAs as compensation for the additional financial cost of conservation management on the farms (CCN, 1986).

Helliwell (1973) produced a method for assigning financial values for a number of British plants and animals. The system was based on index values of abundance, conspicuousness and material value. A series of 'shadow prices' were fixed and by rounding up the figures a value of £6,000,000,000 was placed on British plants and animals. Spellerberg (1980) concluded that while the fixing of financial value may be of relevance in some situations it remains preferable to evaluate wildlife on the basis of ecological and biological valuations.

Ecological Evaluation of Species

The need for sound ecological and biological evaluative techniques has arisen because of the greater pressures now being placed upon all of our biotic resources. Some resources have proved highly vulnerable to pressure (those with very specific habitat requirements or with very slow renewal rates) while species with a wider tolerance of change appear less prone to destruction.

Identification of the very vulnerable species is not too difficult as their reduction in viability is quickly recognised. Problems exist with species which were formerly widespread in extent but which now have become local in their distribution pattern. We must be able to compare former and present distribution patterns, we must explain the reason for change and we must predict the future trend in survivorship . If the trend is undoubtedly towards complete extinction then it will be necessary to alert international conservation agencies and government-sponsored agencies. In order that we may convince governments of the need for action in such circumstances we must have a series of records, for individual species, showing their number, their distribution and their breeding vigour for at least two points in time.

It is fortunate that an international organisation exists to perform this task. Founded in 1934, a Swiss-based organisation called the International Union for the Conservation of Nature and Natural Resources (IUCN) is responsible for recording information on plant and animal species. Data is collected at national level and published in the form of Red Data Books. Coverage is by no means complete, only the higher plants and animals being so far included. Five separate volumes exist:

1. Mammals
2. Birds

3. Amphibians and reptiles
4. Fish
5. Angiosperms - higher plants.

For each species, and each variety in some cases, data on status, geographical distribution, population size, habitat, breeding rate in the wild, conservation measures taken and conservation measures proposed are recorded by scientific compilers. Although the amount of information per species is not great (amounting to about 400 words in total) the data is concisely edited and the use of 'key words' with precise meaning can convey an accurate and abundant account of the status of the plant or animal. Of particular relevance is the information entered under the heading of 'status'. There are five categories of status:

1. Endangered, which applies to species or subspecies most likely to become extinct if current trends persist.

2. Vulnerable organisms, which are severely exploited at the present time or which inhabit areas of major environmental disturbance and which are likely to adapt to those changes. Species in this category will probably move into the endangered status if current trends persist.

3. Rare organisms, which are at risk because of their small, total world population. Rare taxa usually have a very restricted geographical distribution - confined to islands, or mountain summits.

4. Out of danger species, which were formerly in categories 1 - 3, but which have responded to conservation measures and the threat to their survival has been removed.

5. Indeterminate, plant and animals which probably belong to groups 1-3 but insufficient knowledge prevents an adequate assessment of their status.

Some criticism can be made of the subjectiveness of the way in which Red Book data is collected and presented but this is but a small comment to make when set against the enormous value of the data. Any attempt to identify the relative needs of those taxa which require conservation measures to ensure their survival must be welcomed as a positive contribution to wildlife continuation.

ECOLOGICAL EVALUATION OF PLANTS

Red Book data, as might be expected, is most thorough for the higher plants and animals (vascular plants & vertebrates) and best records exist for developed countries. As an example of a detailed assessment of the conservation status of British vascular plants, the British Red Data Book: 1, Vascular Plants, (Perring & Farrell, 1977), can well repay careful inspection. This source book took as its criterion of rarity all plant species which occurred in 15 or fewer of the $10km^2$ grid squares covering Britain. Only post-1930 data was included. A total of 321 species (about 18% of the native British flora) fell into this category of rarity. Forty two taxa had increased their distribution beyond the critical 15 grid square level thus indicating that these species had at least for the time being, successfully avoided the prospective threat of extinction.

Data on the 321 rare British plants was collected by members of the Botanical Society of the British Isles. These members completed a 'Population Form' for each rare species in their area. This form requested a grid reference, a sketch map showing location of the species, size of population, the other members of the plant community in which the rarity occurred, the site conservation status, site use and the site owner. Information from the population form was transferred to an Individual record card, one card being prepared for each $1km^2$ occurrence. In order to remove the subjective elements connected with species rarity the authors of the British Red Book devised a quasi-quantitative Threat Number valuation. This value is the result of the aggregation of a number of index values. The first input reflects the change between present and past distributions, thus:

0 = Decline of <33%

1 = Decline between 33-66% inclusive

2 = Decline >66%

The second input refers to the number of localities (ie 10km^2 squares) at which the species is found:

0 = >16 localities

1 = 10-15 localities

2 = 6-9 localities

3 = 3-5 localities

4 = 1-2 localities

The third input provides a subjective input which refers to the general attractiveness of the species and the associated chance of it being 'collected':

0 = not attractive

1 = moderately attractive

2 = highly attractive

The fourth input is a Conservation Index. It is an arbitrary figure relating to the percentage of the localities of the species which occur within nature reserves.

0 = > 66% of localities in nature reserves

1 = 33-66% localities in nature reserves

2 = < 33% localities in nature reserves

3 = < 33% localities in nature reserves and where these are sites subject to exceptional threat.

The fifth input gives a subjective weighting to the relative ease with which the species can be reached by the public (remoteness value) while the sixth input concerns accessibility of the species when the public have reached the site. Both inputs use the following scale:

0 = not easily reached

1 = moderately easily reached

2 = easily reached

The Threat Number is obtained by adding inputs 1 to 6. Maximum possible value is 15. Of the 321 rare British plants yet assessed, the highest recorded threat value is 13. Eight plant species fall into this category (Table 6.2).

Further examination of Table 6.2 reveals several interesting features:

1. All eight species apart from Pyrus cordata are physically very small and hence easily destroyed.

2. Most of the species could be classified as 'weeds' in that they are associated with man's agricultural systems.

3. Apart form Pyrus cordata all the species occupy sites which have been/are still under considerable modification from man's actions.

4. Apart from one species, none of the species are afforded any conservation measures.

Table 6.2 List of Rarest British Vascular Plants
Threat Number Index 13
data from Perring & Farrell (1977)

Species name	Common name	Habitat	Conservation status
Agrostemma githago	Corncockle	A former widespread weed of cornfields	None
Alyssum alyssoides	Small Alison	Rare annual of grassy fields	None
Damasonium alisma	Star Fruit	Muddy margins of acid ponds	None
Galium spurium	False Cleavers	Arable fields, allotments and waste ground	None
Petrorhagia nanteuilli	Childling Pink	Coastal, sandy gravel sites	1 SSSI
Pyrus cordata	--	Hedgerow shrub in Devon	None
Senecio paludosus	Fen Ragwort	Fen ditches	None
Stachys germanica	Downy Woundwort	Calcareous pasture and roadsides	None

The Threat Number index system is a useful and desirable method of identifying those species most in danger of total disappearance. With modification and development the system could be applied to animals and to areas outside Britain. It should be emphasised, however, that there is a danger inherent in using a Threat Number index system. We could become overconcerned with those species which recorded the highest Threat Numbers. All 321 species identified by Perring and Farrell (1977) are under threat - the 8 species with values of 13 are probably beyond the protection of practical conservation measures. Also, there exist some members of the public who would take great pleasure in collecting species with high Threat Numbers - their 'rarity value' would present a 'collectors value' and possibly a financial value. This could place threatened species under greater stress. Finally, the rarity of a species should not be judged solely on its Threat Number - for example some species may have great sentimental or affection value, or be a national emblem. The giant panda would be an example of the first of these, while the protea plants of South Africa an example of the second.

THE THREAT TO ANIMAL SPECIES

The principal reason for the disappearance of animal species is the same as that for the disappearance of plant species - the alteration of ecosystems in which animals form an integral part.

Animals differ from plants in their reactions to ecosystem and environmental change in that they have the power of conscious locomotion, that is they can move away from unfavourable circumstances to areas which can satisfy their life requirements. Animals also have the ability to adapt to change more readily than plants through their ability to alter their behaviour pattern.

If environmental change is too extreme or occurs too rapidly then even the most adaptable animal will be unable to survive. It will move beyond its normal life zone into the zone of resistance and possibly into the death zone (Fig 6.5).

Animal behaviour patterns are generally more complex than for plants. This is due to the development of an active social relationship with other animals and with plants. Their behaviour is determined both by instinct and by learning and as a result an animal can 'hide' its vulnerability to biosphere change. Allen (1980) quoting from <u>Red Data Book</u> sources has cited loss of habitat (environmental change) as being the major reason for placing animal species on the threatened list. Two-thirds of the species on the threatened list have suffered habitat change. A further 37% have suffered over-exploitation (hunting, persecution and over-kill), 19% have become endangered species as a direct result of introduced 'exotic species', 7% suffer direct competition with man for similar food sources while a small 2% are endangered through accidental killing - for example hedgehogs inadvertently killed whilst crossing roads.

Fig 6.5 Environmental Gradient and Species Abundance

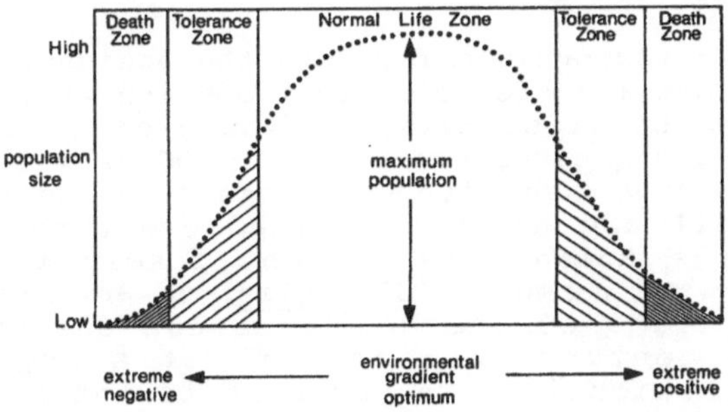

Once again the reason for environmental change and its consequent effect upon animals can be attributed to man's actions. Each time we cut down a forest, plough a grassland, remove a hedgerow, build new houses, roads or airport runways we are destroying the habitats of animals. The loss is not always absolute. Natural forest may make way for a commercial afforestation project and by so doing different birds and mammals will move into the new forest. The raven, crow and kestrel all now nest in British cities, located high on inaccessible cliff like buildings and feeding on the abundant vermin which inhabit our urban areas.

Alteration of Environments

Man is also responsible for the invisible alteration of environments through the massive increase in use of chemical sprays to control pests, weeds and disease. Many of these substances are not totally biodegradable in the environment and accumulate in the food chain. Sometimes they cause spectacular and well publicised disasters amongst animal populations, for example, the Golden Eagle (*Aquila chrysaetos*) in the Highlands of Scotland (*Lockie et al*, 1969). On countless other occasions their accumulation is less well documented and their impact on animal, and plant forms, less well known. For example a recent study of marine pollution (Levy, 1984) has shown widespread surface contamination of oil slicks and floating tar. Dispersal petroleum residues were present almost everywhere to a depth of one metre along all shipping lanes. The significance of this statement is catastrophic when it is realised that the productivity of the plankton in the oceans depends upon the undisturbed purity of the top one to two metres of our oceans.

Man has been responsible for both the accidental and the deliberate transfer of animal species around our planet. The earliest mariners would have carried fleas, ticks and parasites on their bodies while the cargo would have supported countless other animals - mostly small and hence unnoticed creatures. For example, the Romans are thought likely to have introduced the dormouse (*Glis glis*) to Britain. The sea-borne trade of the Mediterranean during the fifteenth century carried with it a huge rat population which in turn was host to the bubonic plague (Howe, 1972).

Islands are particularly prone to the impact of introduced species. This is clearly seen from the avian fauna of New Zealand. A frequently observed trend amongst N.Z. birds has been the loss of flight capability (Falla, 1976). This is particularly so amongst the groups called rails and gallinules; a member of the parrot group - the kakapo - is also flightless. Flight is a useful defence mechanism in that when threatened, the bird can escape. Without the powers of flight the bird becomes confined to the ground, or as in the case of the kakapo, to low growing trees. Survival for flightless birds is only possible when predators are absent. The New Zealand fauna is notable for possessing only two native members of mammals and both of these are bats and none are predatory or flightless birds. The arrival of the first Polynesian settlers in 1100AD and later of the European immigrants in 1826 was accompanied by the introduction of a variety of highly competitive mammals, for example mice, rats, pigs, deer, weasels, polecats, hares, rabbits and hedgehogs (Salmon, 1975). The forest dwellers competed both directly and indirectly for living space and food resources with the indigenous bird life. The list of recent avian extinctions in N.Z. makes gloomy reading; New Zealand thrush, kokako (South Island), huia and laughing owl while others are on the verge of extinction; brown teal, black robin, South Island bush wren, orange-fronted parakeet, kakapo, black stilt, Chatham Island snipe, New Zealand shore plover, New Zealand dotterel, notornis, (Marshall et al, 1977).

International Commerce

The buying and selling of animals, dead or alive, is an important world industry. It tends to be under the control not of the local population, although they can be involved in the catching or shooting of the animals, but is organised by powerful syndicates located by dealers many thousands of kilometres distant. The motivation for trade, much of which is illegal, is financial. Rare and luxurious furs, leather, decorated ivory and exotic meats command massive prices. Other uses for animal products include aphrodisiacs (ground rhino horn), pharmaceutical, investments, museums, zoos and scientific research. Capture of animals as pets can also occur, for example the small song birds illegally captured in Italy. The capture of primates (mainly small monkeys) for use in the bio-medical trade and on a lesser scale for pets, accounts for 160,000-200,000 animals each year. Some of this trade involves cruelty to the involuntary captives. Restaurants in some major cities of the world boast exotic meat dishes on their menus - this is particularly common in S.America, while in some African nations, notably Zaire,

Fig 6.6 International Trade in Rare Animals

TAIPEI: A Bengal tiger destined to be served as gourmet dish at a Chinese banquet has been saved by a rich businessman amid outrage over the slaughter of the animals in Taiwan. The tiger, smuggled in from India to Bangladesh, is now in a zoo after the businessman bought it from a butcher for £7500.

the major source of animal protein (up to 75%) in the human diet originates from wild animals. The newspaper extract (Fig 6.6) shows the commercial value of rare animal species.

Minckley and Deacon (1968), working primarily on fresh-water fish in the S.W. United States of America, have identified four categories of response to environmental change. These groups can be usefully extended to other animals. The categories are:

1. Species which successfully inhabit environments created by or substantially modified by man. This group of species has extended its range and number (European starling, Sturmus vulgaris).

2. Species which do not appear to have been affected by man's actions and which are still still common-place over a wide geographical area (frog).

3. Species which require large, special habitats (golden eagle, Aquila chrysaetos).

4. Species which inhabit small, unique environments, often surviving as relicts or isolated endemics (N.Z. avian fauna).

Animal species in categories 3 and 4 will be most at risk from extinction. Any species, plant or animal, which demands highly specific conditions in order to fulfil its history will find its future existence threatened. Once again, we can only identify man's increasing ability to transform 'wilderness' landscapes into agricultural and or urban/industrial landscapes as the prime cause of environmental change.

The Role of Man

Such is the capacity of modern man to bring about biosphere change that it is possible to reduce a species to the verge or extinction within the space of one or two breeding seasons. This can be achieved either by the physical destruction of a habitat or, increasingly, through the accumulation of pesticides in the ecosystem (Mellanby et al, 1977); Woodwell et al, 1971).

One promising sign for the future of our species is

the fact that we are now able to accept that we, as <u>the</u> dominant species on planet earth, are capable of creating rapid and dramatic environmental change. A corollary of that fact is that as a species, as opposed to individuals within our species, we are gradually recognising the need to exercise restraint on the way in which we treat other plant and animal species with which we share the biosphere.

Exploitation of our biosphere has been a feature of man's behaviour ever since we emerged as a distinct species. Boulding (1966) has likened our attitude towards the biosphere to that of a 'cowboy economy' in which we over-use a resource to a point when its scarcity reduces its usefulness to such a low level that we search for an alternative resource. We then over-exploit it before moving to yet another resource! There is nothing inherently wrong in the way in which mankind (as a species) survives at the expense of other plants and animals in the biosphere. It is simply the fact that mankind is now so dominant, in numerical and capability terms, that if we persist in the uncontrolled use of the biosphere we are likely to bring about detrimental biospheric change. This will come about for the following reasons:

1. Over-utilisation of plants and animals will lead to species extinction.

2. For every one species that we drive to extinction through over-exploitation, up to a further thirty species are transferred from a secure to a vulnerable category because of the interdependency of species within a food web system (Myers, 1985).

3. Extinction of a species will produce ecosystem simplification and ecosystem imbalance (Fig 2.5). As some species are destroyed others will increase due to the new competitive forces. Energy and matter movements will change.

4. Simplification of ecosystems reduces the capacity of the system to persist over time. Feedback mechanisms are disturbed. The ecosystem itself is thrown into jeopardy. It may eventually disappear.

5. As ecosystems disappear then so the circumstances outlined in points 3 and 4 (above)

> will be transferred from the species/ecosystem level to the ecosystem /biosphere level. Biospheric change will occur. This may be beneficial change for man or it may be to our disadvantage.

Because it is impossible to forecast the impact of biospheric simplification on mankind, there is a strong school of thought which has argued that we should make every attempt to minimise further changes to the biosphere (Cailet et al, 1971; Barbour, 1973; Nicholson, 1973).

However, we meet considerable resistance to this idea from the industrialist and economist and also from some politicians. Dasmann et al (1973) have argued that conservation and economic motivation should ideally be directed towards the same point, that is , the optimum use of biospheric resources in order to attain the best conditions for man. Conflict has developed between the conservationist and industrialist because neither group is fully aware of each others aims, motivations and objectives. If the industrialist and developer are to recognise the arguments in favour of conservation then they must be presented with information in a form which they can under-stand. This will probably involve the placement of a financial value on species (see p. 154), along with information about numbers, breeding capacity and an indication of the role played by species within the ecosystem.

In order that we can provide this data we must build upon data held in the Red Books (see p.157). All species, not only the rare species, must be studied. It will involve working at the 'white-box' level of ecosystem study (see pp. 34-35), a task of monumental challenge to ecologists.

Still further effort will be required to construct arguments which can encourage industrialists to provide financial contributions for ecosystem protection. A charge for conservation of ecosystems and species should be included as an overhead on industrial production cost. Not only the industrialist must be persuaded of this argument. A conservation charge will inevitably be reflected in an increase in cost to the consumer. Education and publicity for the need for conservation would be a basic requirement.

Neglected Areas

The pioneering work of the IUCN has already been referred to on page 157. By their own admission, the work of the IUCN has been concentrated on the largest and most visible categories of animals and plants. Very little work has been done on the single largest group of animals to be found on the planet - that of the phylum Arthropoda. Moran et al (1980) has suggested that some 75% of all the animals which inhabit the biosphere belong to this group. It includes the Arachnids (scorpions, spiders, mites and ticks), Crustaceans (comprising aquatic organisms of some 35,000 different species such as shrimps and crabs) and the Insecta class which is the largest class of life forms on this planet. Some 800,000 different insect species have been already identified and there may be a further 400,000 insects still to be classified! Many Arthropods will have become extinct in recent times without ever having been seen by man.

Arthropods have been very successful life forms when judged in sheer numerical abundance. They have also been successful in their capacity to fill almost every possible habitat which exists on this planet. It is possible to suggest that the extinction of a species belonging to the Arthropoda and also the Mollusca phylla is ecologically of less significance than the disappearance of a member of the fish, bird or mammal groups. With such a diverse range of life forms and with the obvious capacity to genetically adapt to a whole range of environments, the loss of a member of the Arthropoda or Mollusca genus would probably be quickly replaced by the process of natural evolution.

The higher animals - notably the vertebrates, are fewer in species number, larger in size and are generally considered to be both more competitive with man for available food resources, and at the same time form a suitable supply of food for mankind. These organisms are thus pursued by man because they take food which would otherwise be consumed by ourselves and also because these organisms provide us with a food supply.

Ziswiler (1971) has produced statistics to show the probable trend in extinction for birds and mammals since 1650AD. His data shows that over 40 mammal taxa and almost 50 bird taxa are currently disappearing every 50 years, (Fig 6.7).

Fig 6.7 Rate of Extinction for Mammals and Birds

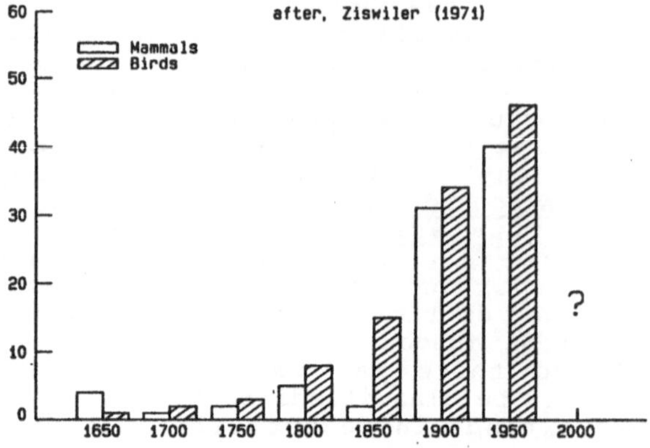

It might be expected that as mammals and birds are relatively large species and are comparatively well documented, compared with the invertebrates, that a well established procedure would be available whereby we could accurately tell which of the higher animals were under threat. Unfortunately, this is not so. Data for the densely populated, industrialised nations of the northern hemisphere is available but these countries, such as Britain, Holland and Sweden, do not have a major problem with animal extinctions. In these countries the massive loss of animal species occurred between 1400 and about 1800, e.g. loss of the bear, wolf and wild boar. Nowadays, most highly developed countries recognised the need for and practise a sound conservation policy.

The situation in developing nations is very different. Conservation costs money and a country with a high balance of payments deficit would be unlikely to embark upon even a minor conservation policy in which the benefits are intangible and at best might not become apparent for a decade or more into the future. Vertebrate extinctions in southern America, Africa and Indo-China are undoubtedly taking place at considerable pace. It is impossible for either official government or independent organisations such as the World Wildlife Organisation to make anything more than semi-quantitative assessments of animal extinctions in the developing nations.

As an alternative it is possible to evaluate the number of species which can exist in different habitats. Small research areas can be established in which species counts can be made along with an indication of the total biological productiveness (the biomass) of the habitat. These small research areas become 'datum' points for information on species diversity, number and productivity and these values can be used to predict the animal numbers and variability over much wider areas and for which general habitat conditions are known. It is much easier and cheaper to detect habitat conditions from air photographs, for example, than to conduct an exhaustive survey of animal numbers and types. The evaluation of habitats as a means for assessing the animal population characteristics for an area is a well established technique.

One of the most famous ecologists of recent times, Charles Elton, has developed a means of classifying habitat types (Elton, 1966). The method was based on the classification of the major type (terrestrial or aquatic), the formation type (woodland, scrub, woody plants, non-woody and open ground) the complexity level of layering (canopy, shrub, field and open layers) and lastly, the use of 'qualifiers' (deciduous, mixed or coniferous). In addition to this assessment of major habitat types it was also found possible to classify the minor habitat types which existed within the major habitats. A total of 34 minor habitats were identified all of which were judged to be important for supporting specific animal life forms. Examples of these minor habitats included dead and dying timber, walls, carrion, and bare soil.

Many other attempts have been made to classify habitats (Bunce & Shaw, 1973; Tans, 1974; Gehlbach, 1975). While each technique tends to favour individual conservation issues, all methods have ensured a systematic assessment of animal diversity. Standardisation of methodology would help allow a comparison between widely different animals located in widely separated geographical regions.

Can Animal Extinctions be Prevented?

The disappearance of animal species is a phenomena which has become apparent only in the last 10,000 years or so of our planet's history. Its cause is clearly due to the ever increasing planetary activities of man.

Extinction of animals can be due to several processes:

1. Hunting a) for sport
   b) to exterminate pests.

2. Destruction of habitat

   a) by deliberate clearance
   b) by unintentional accumulation of pollutants.

3. Competitive forces: species seen to compete for food or space in direct conflict with man will be removed.

4. Disease: species which are vectors of disease will be exterminated.

It would be difficult to prevent the elimination of species under the headings of Competition and Disease. We are committed to the improvement of living conditions for mankind.

It is probably the second heading in the above list which now attracts most concern. Animals cannot live in isolation from their habitat and it is the constant alteration or complete removal of habitat which is now threatening the existence of animals (Allen, 1980).

Can We Prevent Further Species Extinctions?

Given the inherent behavioural traits of our own species it is extremely unlikely that sufficient changes can be made in our life styles which would, in turn, permit the coexistence of all other species. There are very real fears that we may even make our own species extinct if we resort to the use of the ultimate weapon - the nuclear bomb.

The future chances for species survival is by no means entirely without hope. Conservation of ecological resources is still a new activity for mankind, dating back at most a century or so. Already, large areas of our land masses have been placed into conservation care; zoos have taken on the role of breeding rare species and releasing them back to the wild, while legislation has been passed to support and encourage the conservation of ecosystems and species. Perhaps of even greater significance, we now realise that if we desire the prospect of a stable future resource base on which to support our own species, then it is necessary to become less exploitive of resource use and to move towards re-use and conservation of biospheric components.

Undoubtedly, education of the public on the need for a dynamic conservation policy will be necessary. Conservation costs money, and the further along the conservation pathway society travels then the greater those costs become. The public will need to be reassured that conservation expenditure is an essential requirement.

It is difficult to persuade sceptics that we must spend money now on conservation projects the results of which will not be seen for several decades. What is possible, however, is to look back ten or twenty years and to calculate how much expenditure we could save today had conservation measures been taken at that time.

Had we been sufficiently far sighted the problems of acid rain could have been avoided, the eutrophication of lakes largely prevented, the accumulation of chemical residues in soils and water reduced. Almost without exception, ecological scientists have been able to forecast the environmental consequences of specific human activities.

Much of the blame for the disappearance of species and habitats must be placed at the doors of politicians, industrialists and economists all of whom are intent on maximising profits according to the arguments of Keynsian economic theory.

Only gradually are some politicians beginning to realise that many of our biospheric resources have values other than financial. As our technological skills continue to improve then so it becomes possible to use the genetic resources of plants and animals for the service of man. Other organisms can provide medicinal extracts which improve our own survival chances (see Appendices A and B).

If for no other selfish reason than hoping that our own survival can be enhanced through the use of other species, conservation of plants, animals and habitats assumes a new significance.

Chapter Seven

THE CONSERVATION OF ECOSYSTEMS

The Impact of Man on Ecosystems

The main impact of man on ecosystems can be summarised in one word, simplification. This process has come about for a number of reasons:

1. Removal of competitive species; usually confined to larger mammals and dominant plants (trees). Competition can be in the form of:
    a) competition for food resources, and/or
    b) competition for living space.

2. Clearance of an area to make way for agriculture.

3. Removal of species which are antagonistic towards man; e.g. pests, weeds, carriers of disease.

4. Involuntary processes which bring about additional simplification as a result of conditions 1-3.

Deliberate removal of a dominant species can alter the meso- and micro-habitat conditions such that dependent subordinate species can no longer survive the changed circumstances. Allen (1980) has suggested that for every deliberate removal of a major species, the survival of as many as 30 minor species may be thrown into jeopardy.

Clearance of natural ecosystems to make way for agriculture land use is undoubtedly the single most important reason for ecosystem simplification. Prior to the emergence of organised settled agriculture, mankind had little impact on ecosystems. Hunter-gathering communities create few problems for ecosystems because human pressures are light (as little as one person per $20km^2$ (Schwanitz, 1967) and sporadic due to the inevitable nomadic life style of the small population.

Settled agriculture creates very many additional problems for ecosystem survival. The early agriculturalist selected specific sites at which agriculture was possible using only the lowest forms of technology. Thus, light sandy or gravelly soils could be cultivated with a digging stick. Sparsely forested ground would also present suitable sites for clearance using either a stone axe or through the use of fire.

The early forms of agriculture were characterised by 'shifting cultivation' or 'slash and burn' systems in which crops were repeatedly grown at a site until yield fell away due to soil exhaustion and/or erosion. A new site would then be brought into use. This system was very effective in opening out clearings in natural vegetation.

Settled agriculture was far more productive in its food productivity capacity than hunting-gathering societies (Duckham et al, 1970). Population densities increased to about 5 persons per km^2. An increasing population required more food which meant the clearance of yet more land for agriculture. This is a trend which has continued unabated throughout the world up to the present.

Simple Agricultural Systems

Our agricultural ecosystems have been deliberately designed as simple, two- or three- step chains in order that:

a) the system can be easily controlled, and

b) that energy and material transfer can be maximised in a direction which is of greatest use to mankind (Jones, 1979).

In spite of this simplification, a grass field grazed by cattle or sheep, can transfer only about 20% of the potential energy into the agricultural food chain. The remainder is consumed by wild herbivores and by decomposers in the detritus chain.

In order to divert as much energy as possible into a form usable by man, agriculturalists have resorted to extensive use of chemical sprays which eliminate the non-agricultural populations. Unfortunately, many of these chemical additives drift in the air currents onto pockets of non-agricultural land while yet others 'leak' out of the agricultural system into natural systems. Leakage is due to wild animals feeding on agricultural crops and returning to the safety of natural habitats where their excrement and eventually their dead bodies release the chemicals into the soil. The chemicals are then absorbed by wild plants which are eaten by yet other animals thereby transferring the chemicals to additional sections of natural ecosystems (Pimental, 1981).

Ecosystems at Risk

Such is man's dominance on this planet that it is true to say that all ecosystems and every single plant and animal come under some threat of survival. Not all species are under serious risk of extinction (see Chapter Six) but as their habitats and the ecosystems of which they form part become further reduced in size and simplified in structure then so the threat of extinction becomes ever more serious.

The early days of organised conservation (from about mid-1850 to about the late 1930s) tended to be concerned more with the protection of specific species which were under threat (see p. 3). This interest in endangered species is still apparent as witnessed by 'Save the Panda' stickers on the rear windows of cars! Some species appear to generate more than their share of interest from the general public, for example, whales, seals, snow leopard, white rhino, and the Californian condor. Most attention is given to the large, visually spectacular mammals. No campaign has been raised in support of the amoeba, paramecium or spirogyra, all of which used to be common inhabitants of roadside ditches and pools. Vehicle exhaust gases and agricultural chemicals have so polluted the shallow waters of the landscape that the small, unicellular plant and animal organisms have become rarities.

Instead of directing our concern at the disappearance of individual species we should be involved with the conservation of the habitats which support the endangered species. Species loss is the final stage of habitat alteration; species loss is indicative that somewhere and somehow habitats are being altered in ways which make survival of individual species an impossibility.

Concern over the survival of individual species is not conservation, it is in most cases an example of protectionism (see p. 9) in which an exceedingly rare species is given total protection. When a species warrants this intensity of concern then its chances of survival are probably too small and no matter what assistance is given by man, the breeding population has become so restricted that extinction is inevitable (see Appendix C, Californian condor).

Because of the widespread use of chemicals in agriculture and because of the mobility of those chemicals and their ability to spread quickly to non-agricultural areas, all parts of our biosphere are now contaminated with a wide variety of chemical materials. DDT is probably the most widely found artificial chemical, it being recorded in the bodies of penguins from the Antarctic and even from fish dredged from the ocean deeps (Hill, 1982; Regenstein, 1983). World pesticide sales have increased from US$ 8.1 billion in 1972 to US$ 12.8 billion in 1983. Developing countries have shown the most rapid increase in pesticide usage and now account for 15% (1985 data) of the world consumption compared with 8% in 1980. Many of the substances used by developing nations have been banned by developed countries because of their hazardous, cumulative or otherwise unknown effects on the biosphere (World Resources Inst., 1986).

It is difficult, if indeed at all possible, to restrict chemicals to the location at which they were applied. Leakage and drift are inevitable due to water and wind movement and to the transfer of materials through food chains. The transfer of undesirable chemical materials into conservation areas is a scourge for successful maintenance of natural ecosystems. At best it is possible to legislate for a buffer zone surrounding the conservation area in which certain chemicals can be used sparingly and under carefully controlled conditions.

Because of the mobility of chemical pollutants along

with a whole range of other human activities, for example deforesting, drainage, genetic engineering of plant and animal species, the survival of individual species and their immediate habitats cannot be guaranteed. These changes which occur in the name of progress cannot be easily halted.

Of all the vegetation regions shown in Fig 7.1 those which have suffered most alteration are the countries which have experienced the longest occupation by man. Thus the high plateau savanna grasslands of east Africa, the coastal fringes and islands of the Mediterranean Basin and much of central and western Europe have contemporary ecosystems far removed from their original composition. Eastern and central North America likewise have very different ecosystems today than those which existed prior to 1750 AD. Smaller, but no less spectacular changes have occurred on many Pacific Islands. New Zealand shows a remarkable contrast between the grossly over simplified ecosystems of the North Island and the Canterbury Plains and the almost totally virgin ecosystems of the Fjordland National Park at the southern tip of the South Island (see pp. 137-140).

ECOSYSTEMS AT RISK IN THE TROPICS

Within the tropics can be found some of the most diverse, most productive and complex ecosystems which have ever existed in our biosphere (Mabberley, 1983). We are mostly ignorant of the detail of these ecosystems. Early texts describe them as monotonous, drab, devoid of animal life and of little relevance to mankind. Until the early and mid-1960s so little was known of the structure of these ecosystems that Eyre (1970) was prompted to write of the plant families that "they are remarkably similar in structure and general appearance".

While the outward appearance of tropical or equatorial ecosystems may be that of an evergreen, woody dominated system, much of the early work was completed on the fringes of tropical ecosystems, along river courses, or near settlements where gross simplification of ecosystems had occurred.

In reality tropical ecosystems are characterised by maximum diversity and variation. Nowhere else in our biosphere can such complexity in life forms be

Fig 7.1 Distribution of the Main World Vegetation Types. After Riley and Young, (1968)

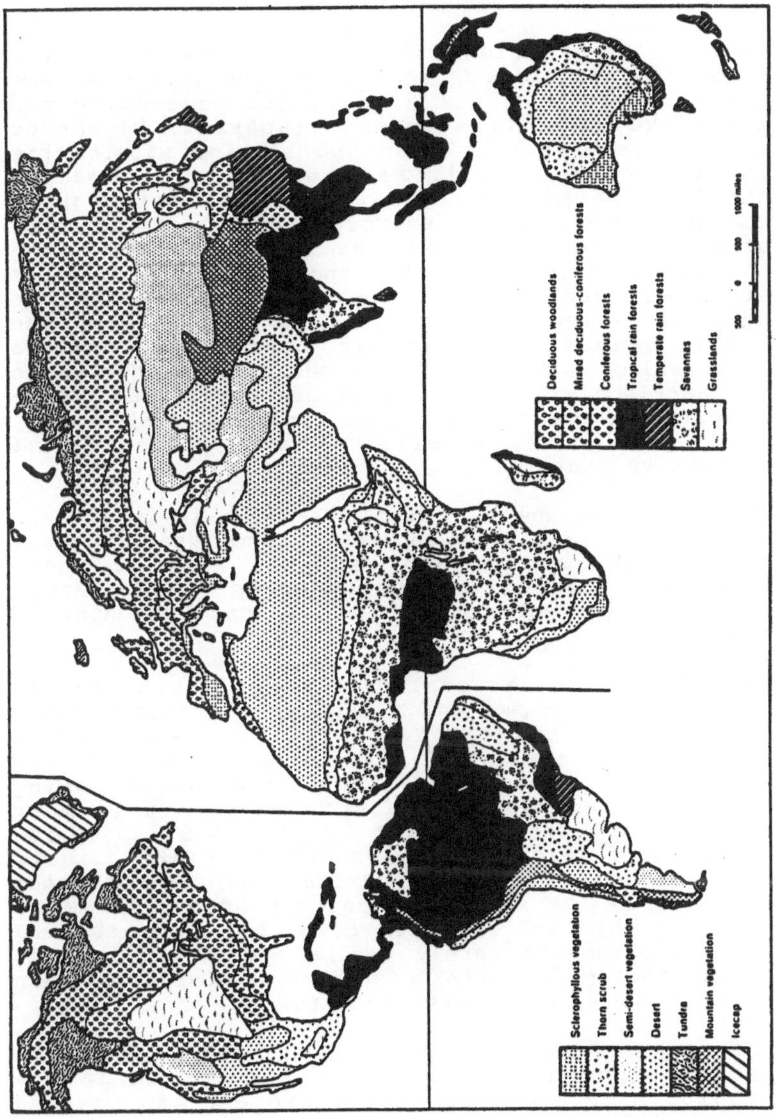

found. Species variety and number reaches its maximum development. Rarely are there fewer than 20 different tree species per $100m^2$ and often the number exceeds 50 and exceptionally reaches 100 different species per hectare (Walter, 1973).

Most of the species within a tropical ecosystem belong to one family, the Dipterocarpacea - a name which refers to the tree's fruits possessing two wing-like structures to aid dispersal. In Borneo alone there are more than 500 species within this family (Myers, 1984). Many of the tree species attain very great heights, 30 metres being quite common, and the dense, evergreen foliage filters out between 90-99% of the outside radiation thus making the forest interior a dark and apparently lifeless area. But many forest floor species have become adapted to low light conditions and, indeed, the space beneath the forest canopy presents a favourable habitat for those which have adapted to it (Harrison, 1962). Evenness of humidity and temperature levels and lack of exposure enable many plants to develop with a marked absence of protective features.

The forest canopy is a whole habitat in itself and supports myriads of epiphytes, saprophytes, lianes, tree ferns, mosses and lichens. Many of these species are rooted on the branches of the canopy layer but project down into the space beneath the canopy.

Animal life is of equal diversity, again divided between those forms which inhabit the forest floor and those of the canopy layer. The relationship between plant and animal life forms is intimate; the evolution of plant and animal communities being interwoven across a time span extending to many millions of years (Gilbert et al, 1975). Richards (1970) has described the complexity of activity which occurs in a South American rain forest; each species has a specific period of activity so arranged that species with competing behaviour and life requirements rarely become active at the same times. MacKinnon (1972) described a similar pattern from the forests of north-east Borneo.

The physical environmental inputs which combine to provide such a unique growth medium within the tropics have been such that European man has had major difficulties in coping with the extremes of heat, moisture and the ecological profusion which results from these inputs. Tropical diseases and the speed at which decomposition processes occur are also complicating factors for the human intruder from extra-tropical latitudes.

More than anything, however, the European 'settler' when attempting to colonise the tropics quickly discovered that northern and mid-latitude techniques of agriculture, building construction, medicine and social organisations were poorly suited to low latitude situations. The simplification of ecosystems which had proved relatively simple to initiate and operate in Europe would not work within the tropics. The profusion of life, the speed of material and energy turnover and the propensity of physical erosion in tropical soils ensured that the European settler was faced with very different challenges from elsewhere on our planet.

Economic Attractions Within the Tropics

The immense richness of the natural ecosystems of the tropics has lured Europeans for generations. Until recently the impact of commercial exploitation on the tropical ecosystems was insignificant; some 400×10^6 ha of land is covered by tropical forest, the largest single category of life forms to be found on this planet. Man could hardly penetrate the vastness of this area.

During the 1960s it became evident that the traditional resource bases of the northern hemisphere would soon become exhausted. New lands would be required to provide timber and agricultural produce for the rapidly growing populations of the northern hemisphere. No new continents remained to be discovered and it was inevitable that the politician and the commercial entrepreneur should turn to the tropics as an alternative supplier of resources.

The profusion of growth suggested that tropical ecosystems were immensely productive. Research by Bazilivich _et al_ (1971) and Whittaker _et al_ (1973) have shown that in terms of biomass and net primary productivity the tropical rainforest out-performs all other ecosystems. Figures of 450 tonnes/ha for biomass and an average $2000 g/m^2$ for n.p.p. have been calculated for numerous tropical sites (Jones, 1979).

Often overlooked, however, is the completely different way in which tropical ecosystems hold their nutrients. They are unique in that nutrients are stored in the living portions of the ecosystem (in the leaves, branches, trunks and root systems) and not in the clay-humus complex of the soil as in all extra-terrestrial ecosystems. Klinge (1975) recorded 97.4% of the biomass of Amazonian rainforest to comprise woody tissue. It is within this material that the nutrient reservoir is to be found.

Repeatedly, however, the commercial farmer and forester, trained in northern hemisphere practice and techniques, believed that one only had to clear away the natural vegetation, plant agricultural crops or trees, wait a remarkably short time and harvest massive quantities of highly valuable products.

In reality this could never happen. As soon as the natural vegetation was removed the nutrient reservoir was rapidly released due to almost instantaneous decomposition brought about by the high temperatures and high humidity. The nutrients could not be stored in the soil as a clay-humus complex could not exist (soil humus underwent bio-degradation in the space of 36-48 hours) and with the ever-present high rainfall more than 2000mm p.a. (Whitmore, 1983) nutrients were rapidly washed out of the soil (Goodland et al, 1975).

Conventional mid-latitude land use techniques will not operate successfully in low latitudes. Removal of the natural vegetation, followed by ploughing of the soil and the planting of a crop will show initial promise but as the nutrients escape from the system then so growth becomes very slow. Another problem is the invasion of adjacent species and unless constant removal of unwanted species is carried out the whole area will quickly regenerate into a thicket of scrub.

Tropical ecosystems have vast wealth in their own right. While it is very difficult to place financial value on biotic uniqueness, the tropical ecosystem is so very different from all others on this planet that it must be accredited with a greater value no matter what currency of assessment is used.

It is possible to place a financial value on

tropical ecosystem resources. Schultes (1980) has recorded 1300 plant species of use as medicinal cures by the indians of north-west Amazonia while Perry (1980) has recorded 6500 species of medicinal value in Southeast Asia. While many of these plants are used only by the indigenous population an increasing proportion are now being incorporated into western pharmaceutical products (see Appendix A). One in four western medicines now contains at least one ingredient obtained from tropical ecosystems and the total value of these substances exceeds US$20 billion per year (Farnsworth, 1982).

Tropical ecosystems can supply just about every possible substance for the food-stuffs, chemical processing and medicinal industries. Tropical plants can be used directly with a minimum of processing, or their chemical and genetic structures can be used as 'blueprints' for the design of synthetic substitutes or thirdly they can be used for research purposes in which totally undiscovered substances may, one day, provide totally new products for mankind.

Myers (1984) has produced a fascinating account of the value of tropical forests as food sources for mankind. Among the most spectacular are:

1. a caffeine-free coffee berry,

2. a perennial corn plant which eliminates the need for expensive ploughing and re-seeding, and

3. Dioscoreophyllum cumminsii which bears a berry with a sweetness some 3000 times greater than sucrose but with no calorific value.

Great future economic promise is possible for the producer countries of these revolutionary crops provided bio-technological solutions can be found to the problems currently associated with their development.

Animals, too, from the tropics can provide major sources of protein (Ajayi, 1979). Many Central African nations gain approximately 25% of animal protein from wild forest animals (Nigeria and Zaire) while in Cameroon and Liberia 70% of protein is gained from forest animals. Myers (op cit) has calculated that a carefully managed forest of some 500km^2 could yield US$10 million per year of animal-derived protein. This is about $200 per ha and compares very favourably with the average return from timber removal which yields $150 per ha and is, at best, a once in twenty years yield.

Forest Management in Tropical Ecosystems

The forestry industry as operated within the tropics is one of the least efficient and most damaging of operations ever put into practice by man anywhere on this planet. Tropical forestry is exploitive, destructive and totally unjustifiable. It is practised by companies who want to make a quick profit and it is encouraged by the governments of producer nations desperate to earn foreign exchange (Hall et al, 1980).

The tragedy of tropical forestry is that its methods are almost totally unnecessary. With just a modicum of forethought a tropical forest represents a utopian renewable resource. It can yield not only the decorative timbers so sought after by the furniture manufacturers of the industrialised countries but also a host of oils, gums, turpentines, resins and latexes. The financial value of these substances is not small. In 1979, Brazil earned US$21,528,000 from export of essential oils for the perfumery industry alone (Myers, 1984)!

Can We Conserve Tropical Ecosystems?

Our ignorance of tropical ecosystems has meant that we have seriously mistreated this vast biotic resource. We have been overawed by its luxuriance into thinking we can unscrupulously remove resources and expect the forest to regenerate its former state. But the structure of tropical forests has meant that in order to gain access to a single tree of commercial value a further ten are destroyed.

Tropical forest ecosystems are of such phenomenal wealth whether measured in scientific, financial, ethical or aesthetic value that it is essential that as large a proportion as possible can be conserved.

Unlike many other extra-tropical ecosystems there are special considerations to be taken into account in the successful conservation of tropical systems.

Because of the immense variety of tropical forest ecosystems it will be necessary to conserve a larger than usual proportion of these ecosystems. At the very minimum 10% of existing tropical forest ecosystems should be conserved and probably nearer 20% would be necessary to ensure conservation of a wide range of communities (Soule et al, 1980; Frankel et al, 1982).

The individual conserved areas must be large enough to include sufficient representatives of all the main species so that an adequate gene pool remains to provide a variable population. The answer to the the question 'what size of population can be considered an adequate gene pool?' can only be guessed at; different species and different habitats will mean a varying population size. In all probability a viable population of each species requires about 10,000 healthy, adult individuals. A frequency of one individual per four hectares is not uncommon, thus for 10,000 individuals an area of 40,000 ha (400km^2) would become the minimal area for successful tropical ecosystem conservation.

Progress with the conservation of tropical ecosystem has so far been very erratic. In South America, the vast Amazonian forest which extends over five international boundaries (Peru, Ecuador, Bolivia, Brazil and Venezuela) has its own unique problems. Some 110,000km^2 of tropical forest ecosystem has been designated for conservation. Designation does not, however, assure successful conservation.

Brazil has been criticised for its apparent disregard of its responsibilities towards its major share of the Amazon forest. Wetterberg et al (1976) has established the existence of eight distinct ecosystems within the Amazon forest. Each ecosystem contains a significant proportion of 'endemics' (a plant or animal species restricted to a relatively small geographical area or habitat). Unfortunately, the Brazilian government has encouraged the immigration of settlers into many of these remote areas. The Transamazonian highway (Bourne, 1978) has accelerated the removal of much of the tropical forest. Logging, burning, grazing of cattle and widespread, indiscriminate use of chemical sprays has already devastated massive areas of forest. Prance (1978) has suggested that despite these losses there still remain sufficient undisturbed areas which could form 'refuges' for tropical species.

By the mid-1980s some 400,000km^2 of tropical forest ecosystems had been conserved. Apart from South American countries, Thailand and Malaysia in Southeast Asia and Gabon, Congo, Zaire and Cameroon in central west Africa have made major efforts at tropical ecosystem conservation.

Monitoring Tropical Ecosystems

Despite the many improvements in transport and in the ability of scientists to collect and analyse data the tropics remain areas of considerable remoteness. Movement, let alone serious scientific study, in those regions away from highways is very difficult. It thus becomes very difficult to monitor the rate of change taking place within the tropics.

The scientist's favoured method, that of direct observation and experiment, is rarely possible. Air photography is ruled out on grounds of the vast cost of the repetitive sorties which would be necessary to record change.

Fortunately, in recent years a new technique has become available and which has revolutionised the monitoring of change in low latitude areas. This is the new science of satellite surveillance made possible by the American LANDSAT programme. A sequence of four satellites has been launched, the last of which, LANDSAT 4, has a picture resolution size of 30m^2 and photographs every area of our planet every 18 days. The data is transmitted to earth in digital form from which it can be analysed directly by computer and colour map images produced (Taranik, 1985).

Already it has been possible to calculate the annual losses of tropical ecosystems (Table 7.1). Depending upon the definition of tropical ecosystem (Jones, 1979) the annual loss amounts to almost 2% of the biome. The rate of loss will inevitably accelerate and it is possible that by the year 2000 AD there will be no tropical ecosystems remaining apart from small conserved areas amounting to between 1.5 and 3.0 million square kilometres. Table 7.2 provides a country-by-country breakdown of tropical ecosystem losses.

Table 7.1 Annual Loss of Tropical Ecosystems
data from Myers (1984)

Cause	Loss in km^2
Commercial timber removal	45 000
Fuelwood gathering	25 000
Cattle grazing (Latin America)	20 000
Farming operations (minimal estate)	160 000
Total loss per annum	250 000

ECOSYSTEMS AT RISK IN EXTRA-TROPICAL REGIONS

Compared with tropical ecosystems all other natural ecosystems appear simplistic and of limited scientific and commercial value. Comparisons are, of course, relative and within the vast range of extra-tropical ecosystems there are individual ecosystems which are of great value.

One of the reasons for the relative simplicity of ecosystems beyond the tropics is the effect of the extensive and long-term presence of man. This is particularly true for large areas of Europe, the Far East, the Middle East and the Indian sub-continent.

Secondly, extra-tropical ecosystems are genuinely simpler than ecosystems located in the tropics because of the effects of climatic seasonality. As one moves both north and south of latitudes 23.5° the climatic regime assumes distinct wet and dry seasons. Beyond 40°N and S an increasingly cool winter season contributes to ecosystem simplification. Moving further pole-ward, beyond latitude 55°N and S there is a net deficiency in solar energy input over the year and the longer day lengths of summer cannot compensate for the high latitudinal position (Walter, 1973).

Thirdly, extra-tropical ecosystems have had less time in which to develop than have the tropical ecosystems. The latter have existed without major disturbance for at least 60 million years (the end of the Mesozoic era) and possibly for as long as 120 million years. Climatic changes have occurred

Table 7.2 Loss of Tropical Ecosystems by Country data from Myers (1984)

Country	Estimated time of extinction
A. Rapid Removal of Tropical Systems	
Australia, lowland Indonesia & Malaysia, Philippines, Thailand Brazil, Central America, Madagascar, E. Africa, W. Africa.	pre-1990
Bangladesh, India, Melanesia, Sri Lanka, Vietnam, Ecuador	by 1990
Remote Indonesia & Malaysia Colombia	by 2000
B. Remote Removal of Tropical Ecosystems	
Burma, Papua New Guinea, Peru Cameroon	by 2000
C. Areas Currently Showing Slow Change	
French Guiana, Guyana and Surinam Zaire Basin	safe until post-2000

throughout our planet (West, 1968) but their impact has been most acutely felt in the mid and high latitudes of both hemispheres. On occasions, climatic deterioration has resulted in the formation of 'ice ages'. The last of these, the Pleistocene, lasted from about one million years to approximately 10,000 years before present. As a result the soils of mid and high latitudes are relatively modern as are the vegetation and animal populations which inhabit these regions.

The effect of these three factors, the impact of man, climatic seasonality and the youthfulness of the habitats, makes it inevitable that extra tropical ecosystems are far less luxuriant in terms of number and life forms than the tropical ecosystems.

The biomes which exist beyond the tropics can be

divided very simply into two main types, the temperate biomes located between latitude 23°N and S and 50°N and S, and further polewards, 60°N and S, the Tundra Biome. The latter is hardly represented in the southern hemisphere due to the tapering nature of the continents.

Conservation of Temperate Ecosystems

Tropical forests degenerate into a variety of grassland savanna systems the majority of which have been formed through regular burning at the hand of man (Flenley, 1979). Given an absence of burning for as little as five years then woody species can reappear.

Warm Temperate Ecosystems

Beyond the savannas a relatively small area of warm-temperate climate exists in south-east Africa, Australia, South America, the Gulf states of USA, southern China and southern Japan. The climate of these areas are typified by abundant rainfall, hot summers and short cool winters. For the most part, vegetation is evergreen. These areas have proved very suitable for productive agricultural crops which demand a long, frost free growing season (cotton, maize and rice). Very little unmodified warm temperate habitats now remain.

The main area of conservation centres on Australia with its distinctive marsupial animal population and its abundance of Eucalypt tree species. The latter are highly resistant to fire damage and provided fire frequencies do not exceed one in five years the Eucalypt forest can thrive. The forest supports a very large animal population with diverse members of birds (in particular the parrots), insects and the marsupials.

A very complex relationship exists between the eucalyptus forest and the animals it can support. Many eucalypts are poisonous to all but one or two species of animals. It is thus obvious that commercial logging for specific varieties, for example, the Sydney blue gum (*Eucalyptus salignum*) or jarrah (*E. marginata*) will jeopardise the existence of specific animals (Breeden *et al*, 1975).

The island of Tasmania is of extreme conservation value. The Bass Straits which separate it from mainland Australia have been wide enough to allow a distinct Tasmania fauna and flora to develop. Rare mammals such as the Tasmanian devil (Sarcophilus harrisii), a small rat-like creature, still survives although the fate of the Tasmanian wolf (Thylacinus cynocephalus) is less certain.

Conservation of Australian habitats including the warm temperate habitats has been catered for by the establishment of 60 national parks and numerous reserves and sanctuaries. Unfortunately, commercial logging interests have persuaded the Australian government to allow deforestation to occur in many national parks, including most recently (1985) the felling of some of the finest remaining eucalypts in Tasmania.

Mediterranean Regions

The Mediterranean regions consist of five isolated areas: central Chile, California, the extreme southern tip of South Africa, the areas around Perth and Adelaide in Australia, and the coastal fringes of the Mediterranean basin from which the climatic type gets its name.

The five areas are characterised by very hot, dry summers during which all plant growth ceases, and a cool, moist winter which is the growing season. It is the only biome with a 'reversed' growth period (winter growth, summer dormancy). It is also unusual, though not unique, in having a complete admixture of hardwood trees and conifer trees (di Castri et al, 1973).

Mixed forests once clothed the Mediterranean lands but the attractiveness of the climatic regime ensured that man has become concentrated in these areas from very early times, 7000 years BP. Deforestation has proceeded apace and today the visitor to these areas would find it difficult to believe that luxuriant forest would have once grown on the rocky terrain.

Removal of the forest allowed the contrasting effects of the winter rains and the scorching summer sun to quickly erode the soil until, in many areas, bedrock now appears at the surface.

Alteration of Mediterranean ecosystems has been so complete that no representative area remains in the northern hemisphere apart from the tiny island of Port Cros to the south-east of Marseille. This has formed one of the small number of French national parks since 1963. In the southern hemisphere a number of conserved areas exist, for example, the Cape of Good Hope nature reserve with its unique sclerophyllous flora and a further half dozen reserves in Australia. It is unlikely that any other areas worthy of representing this very attractive biome remain in a sufficiently unaltered state to justify conservation status.

Temperate Grassland Biome

It has been estimated by Barnard (1966) that some 25% of the land surface of our planet is now covered by grassland. Apart from some areas of grassland of proven antiquity (parts of the South African Veldt, the South American Pampas and the Steppes of Russia) all other grasslands are relatively young in age and are the direct result of man's agricultural practices.

In natural forest ecosystems, grasses are comparatively rare components of the species mix. Removal of the forest dominants allows a massive increase in radiation to reach ground level and the grass family (Graminaceae) responds rapidly.

Grasses also will tolerate grazing by animals. They possess a genetic structure which adapts quickly to changing conditions and can thus colonise new areas. This last characteristic has been utilised by man and the Graminaceae family has provided all our important grain foods (wheat, barley, oats, rice).

Man has made great use of grasslands, first for grazing cattle and more recently as locations for the cultivation of cereal crops. The latter activity has sometimes been extended into areas of marginal suitability and this has led to massive soil erosion both in the USA in the 1930s and in the Soviet Union in the late 1950s (Silvestrov, 1971).

Substantial conservation work has been undertaken in North America and the Soviet Union to stabilise and restore former grassland areas. This has taken the form of windbreaks, shelterbelts, contour and strip cropping, eradication of weed species and reduction in overgrazing.

Grassland conservation is essentially undertaken for economic resource reasons - that is to reinstate the agricultural productivity. Some old-field grasslands now have a scientific conservation status but these are small in extent.

Such is the demand for good agricultural land that it is unlikely that areas of grasslands will be made available for scientific conservation. This is unfortunate as they can be particularly important sites for the sustenance of small mammals (rabbits, voles, gophers), large grazing herbivores, in particular deer, gazelle, and the carnivores which are dependent upon the grazing animals (polecat, wolves, birds of prey).

Temperate Deciduous Forest Biome

This biome was predominantly to be found in the northern hemisphere in a latitudinal belt from about 40°- 50°. In the southern hemisphere the biome is confined to southern Chile.

Seasonal variations in the climate are well developed with warm summers and often a winter phase of two months when average temperatures may fall as low as 0°C. Moisture availability is rarely a limiting factor. The original vegetation cover of this biome comprised mainly deciduous hardwood species with members of the oak species (*Quercus*) being common. Species variety was good, with up to 20 species per hectare being recorded for Europe, with up to 40 species/ha in N. America (Eyre, 1971).

The forest has formed an important resource base for man from about 5000 years B.P. up to the late 1700s by which time the European forests had been almost completely cleared. Timber for building purposes, for charcoal, for ships, carts, furniture, fencing and utensils all made a rapid toll on the forest. The soils beneath the forest were deep and fertile. The Brown Forest Soil which is characteristic of the deciduous forest has proved capable of continuous cultivation without the problems of erosion which are associated with many other soil types (Russel, 1968).

A complex animal community developed in the forests with distinct species stratification into forest floor or forest canopy dwellers (Elton, 1966). A wide range of animal types occurred including large mammals (wild boar, badgers, bears,) a prolific bird population and an even greater population of decomposer organisms.

Much of the deciduous forest has been cleared and the few areas that remain have been so greatly disturbed by man that the present structure bears little resemblance to its former state (Tubbs, 1974). The woodland remnants of Europe can be classified as belonging to one of the following classes:

1. Sites at which woodlands have existed for a very long time (more than 2000 years).

2. Woodlands comprising native species which have arisen more recently due to a cessation of grazing and cultivation.

3. Woods of native species which have arisen from deliberate planting.

Woodland of very long ancestry can be substantiated only by means of time-consuming pollen analysis of the soils (Dimbleby, 1976). Such sites at best are mere remnants of much larger forests. They have undergone competition and erosion in size from other land uses until in many instances the minimal area is incapable of sustaining a healthy breeding population. Old hedgerows or scattered remnants of woodland can sometimes serve as useful gene pools and form suitable sites from which woodland regeneration can occur.

The second category of woodlands - native species which have recolonised former agricultural sites - are not uncommon. They tend to be located in areas of agricultural marginality and thus can be remote from markets (an advantage in conservation terms) but are also subject to relatively austere climates (a disadvantage). These woodlands are often small in size and confined to steep valley sides. Many of these areas have been given special conservation status in the form of Nature Reserves. Community structure is usually totally different from natural woodlands as the age structure of the trees is uniform. Woodland regeneration began at the time when agricultural pressures ceased. These woodlands now require careful management in order that they follow a succession pattern which will lead them more closely towards the structure of a totally natural woodland.

The third category comprises the planted woodlands. They frequently show uniformity of age, have a restricted species composition and show evidence of former management practice - most commonly the coppice with standards. Sometimes they may also contain exotic species. The older planted woodlands, pre-1914, have usually been stocked with native species while post-1918 plantings have relied upon exotic conifers. The latter are alien to the local environment and cannot support native fauna nor ground flora. They can be justified solely on economic grounds.

Substantial fragmented areas of temperate deciduous woodlands still exist. Although they are small in total area they are of immense scientific, aesthetic and recreational value. They form sanctuaries for many plant and animal species. With careful conservation management and supported by a sympathetic public attitude towards conservation they could form the nuclei for a gradual extension of deciduous woodland. It is unlikely that these areas would ever attain very large size, perhaps extending to about $10km^2$, but their value as conservation areas would be enhanced if they were linked by means of hedgerow or shelterbelt corridors along which migration of species would be possible (see Fig 3.5).

Conservation of Tundra Biomes

Tundra biomes are confined to latitudes beyond 60°N and S of the equator. They are characterised by conditions of extreme environmental hostility with periods of intense cold (winter averages of -10°C being commonplace for between six and ten months of the year). Precipitation is scanty, 250-300mm p.a., much falling as snow (Ives *et al*, 1974).

The bioclimate of tundra areas is made particularly difficult for plant and animal survival due to:

1. The presence of permafrost - the perenially frozen layer which occurs at often just a few centimetres below the surface.

2. The high exposure levels due to the almost continuous blowing of cold, dry winds. The winds dessicate cell tissue while plants rooted in the shallow soil often fail to obtain sufficient

water and hence die of water shortage. Permafrost also creates an unstable surface and any deep rooted plants are disadvantaged by the loosening of the root structure and even by the snapping of deep roots.

Vegetation of the tundra is typified by treeless waste lands in which lichens, mosses, sedges and rushes predominate with small colourful patches of flowering herbs intermixed with small patches of grass species and prostrate tree forms. Plant growth is exceedingly slow and survival is precarious, often relying on the chance protection afforded by snow-beds. If snow melt occurs prior to the start of the growing season then the developing vegetation can be ripped apart by the winds. Plant productivity values for this region are minimal, ranging from $3g/m^2$ per year to a maximum of $140g/m^2$ per year (Jones, 1979).

Animal life is confined to summer migrants. Few animals can survive the winter conditions. Those that remain in the area resort to deep hibernation, for example, the Polar bear (Thalarctos maritimus). In summer a large population of migratory birds arrive to feed off the abundant insect life and the numerous fish which inhabit seas, rivers and lakes. Large herds of herbivores, Caribou (Rangifer tarandus) and the Moose (Alces alces) in North America and the Elk (Alces Alces) and Reindeer (Rangifer tarandus) in Russia and Scandinavia feed off the abundant herbage which grows in summer.

For generations the tundra had supported small populations of highly adapted groups of Eskimos and Lapps. They had developed a life-style based upon fishing and hunting and which was in almost perfect sympathy with the environment (Murdock, 1961). In recent decades their life-styles have undergone a total transformation (Mitchell, 1983). The coming of the geological prospector, soon to be followed by the mining engineer with their mechanised transportation and imported life-styles characteristic of the middle latitude cities has caused a rapid and massive severance of the natural order.

Tundra ecosystems are amongst the most fragile on earth and the effect of disturbance, from whatever source, can be disproportionately large (Bliss, 1971). The discovery of oil and natural gas in Alaska and in Siberia has led to extensive habitat destruction. Pollution by diesel fuel is a particular hazard while tracked vehicles cause extensive damage to both vegetation and soil (Shchelkunova, 1976).

Damage done to any ecosystem in our biosphere is difficult to repair. In the tundra environment repair work is made impossible by the slowness of growth and development. The damage made by a walker's boot to a plant can take up to 30 years to repair!

There have been few attempts to conserve areas of tundra. Two large areas of conservation occur in the USSR, these are the Pechors - Ilych National Park and the Kandalakcha-Lappland National Park, in Finland can be found the Lemmenjoki National Park while in the USA the sole representative of this biome type is the Mount McKinley National Park in Alaska. Apart from these no formal conservation has occurred. As man turns increasingly to the tundra areas for mineral and fossil fuel resources it is inevitable that extensive destruction of these ecosystems will occur.

While tundra systems may not have the scientific appeal of the tropical ecosystems (in that they do not have the range of genetic resources) they nevertheless represent one of nature's last great wilderness zones. Tundra represents ecosystems which have developed the most sympathetic of adaptations to the severely limiting physical environment inputs.

In an increasingly crowded world with competition for space, resources and dumping grounds for the waste products from our industrial societies, it is inevitable that the empty areas of the tundra should be viewed as increasingly useful areas. It would be a tragedy if we forfeited all our tundra to the greed and avarice of man.

Chapter Eight

CONSERVATION AND PLANNING

Throughout the history of mankind, technological advances have been used to further the development of our own species. Unfortunately, while the application of technology may be of benefit for mankind it is usually detrimental for all non-domesticated plants and animals.

The effect of man's technology on the biosphere has brought about the following changes:

1. The removal of species - extinction.

2. The modification and/or destruction of habitats.

3. The degradation of quality in the biosphere through the introduction of pollutants, see Table 8.1.

Deterioration in biospheric conditions has caused a progressive demise in the general health, vigour and reproductive capacity of many species, particularly the vegetation which is forced to endure pollutants by virtue of the fact that they remain rooted in one place.

Ward and Dubos (1972) summarised the opinions of many of the speakers at the U.N. Conference on Environment in Stockholm (1972) in the following statement:

> The two worlds of man - the biosphere of his inheritance, the technosphere of his creation - are out of balance, indeed, potentially in deep conflict. And man is in the middle.

Table 8.1 Summary of Sources, Annual Emission, Background Concentration and Major Sinks of Atmospheric Gaseous Pollutants, after Cheremisinoff & Morresi (1977)

Pollutant	Major Source: Anthropogenic	Major Source: Natural	Estimated Emission (kg): Anthropogenic	Estimated Emission (kg): Natural	Background Concentration (ug/m^3)	Major Identified Sinks
SO_2	Combustion of coal and oil	Volcanoes	65×10^9	2×10^9	1 - 4	Scavenging; chemical reactions, soil & surface water absorption; dry deposition
H_2S	Chemical processes; sewage treatment	Volcanoes; biological decay	3×10^9	100×10^9	0.3	Oxidation to SO_2
N_2O	None	Biological decay	None	590×10^9	460-490	Photodissociation in stratosphere; surface water & soil absorption
NO	Combustion	Bacterial action in soil; photodissociation of N_2O & NO_2	53×10^9 combined with NO_2	768×10^9	0.25-2.5	Oxidation to NO_2
NO_2	Combustion	Bacterial action in soil; oxidation of NO			1.9-2.6	Photochemical reactions oxidation to nitrate; scavenging
NH_3	Coal burning fertilizer; waste treatment	Biological decay	4×10^9	170×10^9	4	Reaction with SO_2; oxidation to nitrate; scavenging
CO	Car exhaust & other combustion processes	Oxidation of methane; photodissociation of CO_2; forest fires; oceans	360×10^9	3000×10^9	100	Soil absorption; chemical oxidation
O_3	None	Tropospheric reactions and transport from stratosphere		(?)	20-60	Photochemical reactions; absorption by land surfaces (soil + veg) and surface water
Nonreactive hydrocarbons	Auto exhaust; combustion of oil	Biological processes in swamps	70×10^9	300×10^9	CH_4=1000 non-CH_4<1	Biological action
Reactive hydrocarbons	Auto exhaust; combustion of oil	Biological processes in forests	27×10^9	175 10^9	<1	Photochemical oxidation

Increasingly, we are placing the responsibility of caring for the biosphere into the hands of the planner and the legislator. By so doing it appears that we accept that mankind cannot voluntarily behave in a responsible fashion towards the biosphere; instead we are prepared to accept a form of compulsion through planning control.

Planning for Conservation

Conservation planning in its strictest sense, in which habitats and species are given total conservation, is comparatively rare. Instead, conservation planning is usually combined with at least one other planning objective, for example, with tourism, recreation, or rehabilitation of derelict sites.

Allocation of land to exclusive conservation use is restricted by:

1. Scarcity of land, water and air resources which, in turn, makes multiple use of biosphere resources inevitable.

2. Conservation is judged to be an expensive land use. There is no recognisable and immediate return in terms of a 'productive harvest' from conservation land use. Instead, conservation planning deals with long-term 'futures' some of which may not reach fruition for many years.

Of the many attempts to improve the commitment to conservation control none can exceed that which exists in the USA. The National Environmental Policy Act of 1969 represents a major bench-mark in the planning and legislative control of environmental deterioration. The Act requires every new federal development to include in its planning application a statement indicating quite clearly how the proposed development would affect the environment (Cheremisinoff _et al_, 1977).

In practice the National Environmental Policy Act has become strangled by legislative red tape. Environmental management has not made the hoped-for breakthrough originally intended by the act. Some ten years after the act passed the House of Representatives, the International Union for the Conservation of Nature made public its _World Conservation Strategy_ (see p.92). This document clearly indicated the depressing fact that such was man's prowess at despoiling the biosphere that all plants and animals along with their habitats could no longer be assured of retaining their self-regenerating capacity (IUCN, 1980).

Legislative control of the environment and its use is only effective when it is correctly applied and enforced. For every legislative action placed upon the statute book there will be an accompanying list of exemptive clauses and conditions, in effect 'loop holes' allowing certain industries and individuals to pay little or no attention to the legislation.

Many of these legitimate escape routes are the result of inexperience in applying strict legal control to a dynamic entity - the biosphere. It is impossible to anticipate all the permutations which might occur in the management of the biosphere and the lawyer can only respond to the expert ecological advice and introduce a series of 'conditional clauses'.

Other instances occur when legislation is deliberately flouted because enforcement of the law does not occur. There are many examples. Motor vehicle exhausts must now meet stringent pollution control levels for CO_2 and the NO group. Many motorists tamper with carburettor settings in an effort to improve fuel economy and in the process violate the pollution emission levels. Few of these culprits are ever penalised.

Conservation legislation is only as good as the ecological advice upon which it is based. The requirement for sound ecological information as a basis for successful conservation legislation throws the challenge back to the ecological scientist. The most useful contribution that the ecologist can make to society is the design of a realistic conservation strategy. The term 'realistic' is emphasised because there is no purpose in advocating an elitist approach to conservation. Elitism is interpreted, wrongly, by the general public as being synonymous with extremism and for the most part extremist views do not make political headway.

A realistic approach to conservation is also necessary because conservation is expensive and the taxpayer has every right to demand an efficient and cost-effective use of taxation revenue.

Practical Considerations

Most ecologists and many conservationists see their role as the scientific study and research of the biosphere. The result of that research is published in learned journals and as conference reports. Much of the information lies hidden on library shelves or on computer tape never to see direct application in the real world.

Of far greater relevance would be ecologists who are capable of providing the background information for planners, politicians, industrialists, economists, architects and a host of other users of environmental resources. What we require is a plentiful supply of well trained ecological planners who have been taught the principles of ecology and of planning and who can translate these principles into practical conservation planning.

The task of the ecological planner is not an easy one. It involves the persuasion of other planners, who may have little or no knowledge of ecological systems, of the need to offer protection from further alteration to all organic components which inhabit an area. Conservation is no longer the preserve of rare species. The situation has been reached whereby entire habitats can be destroyed in the name of economic progress. Complete habitat conservation therefore becomes necessary.

Many examples of habitat destruction can be found in the increasingly commonplace way we build new towns or factories on 'green-field' sites. Motorways cut swathes of unproductive land through often high quality lowland sites. Airport runways have been extended into infilled coastal wetlands as in Auckland and Hong Kong. Sea-port installations searching for flat land for container storage areas also tend to spread in linear fashion across coastal flats and by so doing destroy breeding areas for sea birds or cut off productive algal populations from their required sources of nutrients. It is difficult for the ecologist to argue for the retention of mud-flats which may be used for three or four months of each year for migrant birds when the industrialist and politician may insist that development of these sites would bring additional employment prospects and improved national growth.

In Britain the objectives of the conservation

planner have been thwarted in recent years by an entirely legitimate process, that of the private parliamentary bill. The private bill procedure allows private companies or local authorities to present development plans direct to the House of Commons at parliament. Normally, such plans would be subject to planning procedures, but the private bill circumvents this requirement. Instead of the usual public enquiry a parliamentary committee would judge the merits of the proposed private bill. Statutory bodies are severely restricted in the amount and type of evidence they can present under these circumstances. Significant land use changes have been achieved in recent times using the private bill, for example, the Felixstowe docks extension and the container terminal at Falmouth. The largest private bill for many years involves the Channel Tunnel construction along with associated south coast road developments.

O'Riordan (1979) has claimed that the ecological planner must embark upon a more active role in which research is accompanied by recommendations to the decision makers. If the ecological planner can convince the decision maker that conservation does have a political value (that is, it is a vote-winner) then conservation can compete head-on with the industrial and economic development arguments.

The ecological planner must learn to work on the fore-shortened time scale of the politician if conservation planning is to have a chance of success. Just as the ecologist can justifiably complain that the politician does not understand the ecological argument then so must the ecologist realise that the politician cannot wait for the result of a long-term research project to yield its findings which in themselves may or may not be conclusive.

The speed of social change is constantly increasing. A vote winning issue today becomes an 'also-ran' in the near future. Some issues are hardy perennials - defence, agricultural policy and income taxation. Can the conservation issue become sufficiently politicised that it too can be a continuous vote winner?

There is a real danger that by forcing conservation to the constant forefront of public awareness total 'switch-off' will occur. If the conservationist argument is continually one which advocates 'gloom-

and-doom' unless a radical change in society becomes possible then the traditionally conservative public will soon lose interest in the conservation argument. An approach which advocates imminent catastrophe unless we prevent pollution from thermal power stations, or cease using all chemical sprays on our crops or restrict further expansion of our population size, will only reinforce the general public attitude that conservation is primarily involved with suppression of personal freedom. Instead, conservation issues must be shown to involve an expansion of personal responsibility towards the biosphere. Only when we are prepared to operate in harmony with the biosphere can we expect a genuine and lasting expansion of personal freedom.

The need to provide recommendations on ecological issues in a fast, reliable fashion poses a major problem for the ecological planner. The ecologist would inevitably wish to design a field experiment and compare it with a 'control' site. He would also hope to make a 'before' and 'after' study so that the impact of an event could be genuinely studied. He may also wish to replicate the experiment many times over in order to provide a result which is statistically valid.

An ideal approach to ecological problem solving is rarely possible when confronted with the realities of planning biosphere use.

Many of the problems which the ecological planner would be asked to comment on would be hypothetical 'what if?' questions. For example, what would be the impact on the local environment of a public utility waste incinerator plant which had a predicted pollutant output of 20 tonnes water vapour per hour, 1 tonne fly ash per hour, 0.25 tonnes CO_2 per hour and SO_2 air pollution levels of 25 parts per million (p.p.m.) and NO_2 of 2 p.p.m.?

How could a response to this type of question be given? It would be impossible to conduct a field experiment due to lack of time and financial resources. Instead, the ecological planner would be best advised to look for an existing waste disposal plant in which similar design and operating principles existed. Provided that as many of the control variables between the 'real' and the 'hypothetical' situations could be made constant then the impact of the hypothetical waste disposal plant on the local environment could be inferred.

For most of the time the ecological planner would be involved with small-scale issues - an extension to a sewage treatment plant or a new housing development. From time to time major development projects would arise, for example, a major new highway, a complete new town, a petro-chemical factory, or most controversial of all, a nuclear power station.

The large-scale development is usually accompanied by a public enquiry which is often of a quasi-judicial nature. There would undoubtedly be extensive press, radio and TV publicity in which the ecological planner would be expected to take an active part. Many ecologists have received little, or no training for this role of publicity agent and confronted by a carefully orchestrated and lavishly produced counter-argument from the promoters of the development the ecological argument can appear lack-lustre and even contradictory (Selman, 1981).

In this situation the ecologist would now find himself in the role of 'expert adviser'. He would be cross-examined by the supporters of the proposed development. Any weaknesses in the ecological argument would be exploited and

> when ecological investigations are incomplete and/or the processes are too complex to assure unambiguous findings, vested interests can define the problem and pinpoint solutions to suit their own arguments. It is . . . possible for the serious ecological researcher to get caught in the cross fire and to find his results interpreted to suit particular prejudices.
> O'Riordan (1979)

APPROACHES TO CONSERVATION PLANNING

Kain (1981) has claimed that conservation planning is no longer the preserve of "enlightened individuals or altruistic groups as its champions". Today it stands as an integral part of environmental and resource management and forms an essential component of the planning infrastructure of all developed and many developing nations (Conant et al, 1983).

The National Environmental Policy Act

The passage of the National Environmental Policy Act (NEPA) of 1969 by the American Congress can be viewed as the most significant piece of conservation legislation that has been enacted this century (House et al, 1977). That is a bold claim; can it be authenticated?

There have been other very important acts passed by other countries and which have set in motion a world-wide trend in support of conservation issues. The Clean Air Act passed by the British parliament in 1956 was of monumental significance in ridding the atmosphere of its particulate matter (grits, ash, dust and soot). It has been adopted by most other industrial nations. Unlike NEPA, it was concerned with only one aspect of environmental concern, namely that of air pollution.

In the USA, acts such as the Wilderness Act (1964) and the Wild and Scenic Rivers Act (1968) were all major contributions on specific conservation issues which have also been copied by many other nations.

NEPA differed from other conservation legislation in that it was the first to recognise a number of biosphere problems all of which had a common theme - the problems were the result of man's own actions. It is worth examining the four basic tenets of the act. They are:

1. The declaration of a national policy that will encourage productive and enjoyable harmony between man and environment.

2. The prevention or elimination of damage done to the environment while stimulating the health and welfare of man.

3. The creation of interest and understanding of our nation's natural resources and ecosystems.

4. The establishment of a Council on Environmental Quality.
(Cheremisinoff et al, 1977).

The act represents an acceptance, on behalf of a national government, that the actions of man upon the biosphere can be of a damaging nature to the very integrity and survival of plant and animal species, habitats and environment.

NEPA is important not only because of its acceptance of man's total responsibility for biosphere degradation but the act is a bench-mark in the evolution of conservation ideology and its incorporation into government policy. The politician in the USA has recognised that conservation issues can win votes. A similar situation also exists in some other highly developed nations. Sweden, New Zealand, Australia and some of the Canadian states have also passed legislation similar to that contained within NEPA.

Important though NEPA may be as an official government statement, the act is given practical significance through the provision of a clause which requires the publication of an environmental impact statement (EIS) to determine, before implimentation, the environmental effects of a proposed action.

Thus any major federal action which requires planning permission must include an assessment of the effect that action might have upon the environment. Ideally, the assessment would include an impartial account of any adverse environmental effects which could not be avoided should it be decided to proceed with the development. It should also include alternative recommendations which might reduce the effect of the development on the environment but by so doing might impose greater financial costs for the proposed development.

A prime objective of EIS is for the developer to consider long-term effects of the action on the environment and for these effects to be given greater weight than short-term gain.

An EIS is not a legally binding document. It cannot commit a developer to any specific course of action but becomes part of the total data whereby the decision maker can reach the optimum conclusion.

A shortcoming of NEPA legislation was that it applied only to developments involving federal finance or federal licensing or permits. Thus, a great many private developments escaped the control of NEPA. However, most of the major US companies are very concious of their overall public image. Involvement and concern with the environment is perceived to offer high returns in the form of both government and public approval. Thus, most companies spend large sums of money on conservation, either through the employment of their own ecological advisers or else through private consultants. Admittedly, much of the money is spent on developing a conservation 'image' or on conservation 'cosmetics' in which amenity or landscaping considerations play a major role. Equally real, however, is the trend towards genuine long-term conservation issues, for example, west coast logging companies developing techniques which minimise the disturbance to those forest areas which are not to be felled.

Table 8.2 provides a summary of the areas of environmental concern which are covered by NEPA legislation. It can be seen that the scope of the act is very wide. Because of the breadth of interest covered by EIS control, the operation of the system has become burdened with bureaucratic control, high costs and long delays. In extreme cases, complex EIS reports can delay a project for up to two years. The preparation of an EIS report normally takes between nine and twelve months adding considerably to the cost of a development.

Some criticism of EIS has been due to the less than adequate means by which EIS has been integrated with the planning process. Environmental assessments must be incorporated into a project at the earliest possible stage and should not be included as an afterthought. Not all the delays have been due to inadequacies in EIS itself. Developers have caused many delays through the deliberate challenging and blocking of EIS schemes via the law courts.

Achievements of EIS

NEPA legislation has been criticised by conservationists and developers for, on the one hand, its lack of real conservation achievement and on the other, its cost and delaying abilities. It is relevant to pose the question 'what would be the state of the environment if the National Environment Policy Act had not become law in America on 1 January, 1970?'

Here lies a major and still to be solved issue for the conservationist and ecological planner. How can we assess 'what might have been' if another course of environmental management had been selected? It is

TABLE 8.2 Areas of Environmental Concern to NEPA Legislation after Cheremisinoff (1977)

Atmosphere	Hazardous Substances	Land Use and Management
Air Quality	Radiation	Land Use Change, Planning
Weather Modification	Toxic Materials	Planning and Regulation of Development
	Pesticides	Public Land Management
	Transporting and Handling Hazardous Materials	Protection of Environmentally Critical Areas
Energy Supply	**Hydrosphere**	Land Use in Coastal Areas
Electricity Generation and Transmission	Water Quality	Redevelopment in Built up areas
Petroleum Development Extraction, Refining, Transport and use	Marine Pollution	Density and Congestion Mitigation
Natural Gas Development, Production, Transmission, and use	Fishery Conservation	Neighbourhood Character
Coal and Mineral Development Mining, Conversion, Transport and use	Shellfish Sanitation	Impact on Low Income Population
Renewable Resource Development	Waterway Regulation	Historic, Architectural and Archaeological Preservation
Energy and Natural Resources Conservation	Stream Modification	Soil and Plant Conservation
	Fish and Wildlife	Outdoor Recreation
	Solid Wastes	
	Miscellaneous	
	Noise	

an example of the Catch-22 situation; if a course of action for environmental management is acted upon then the ecologist will be judged upon the results of the proposal. Any delay in proposing an environmental management plan in fear of making an incorrect proposal will only lead to criticism from other sections of the public.

Provided that a recommendation is based upon sound ecological experience and judgement then it can be argued that action will always be better than delay or indecision.

Environmental Impact Analysis (EIA) techniques have been developed which allows a wide variety of environmental problems and areas to be evaluated (see Table 8.3).

European Involvement with EIA Schemes

The European Economic Community (EEC) has been slow to come to terms with a standard approach for EIA schemes. The West German Federal Government made the earliest attempts (1971) to integrate EIA into its existing planning framework. It is interesting to note that the Green Party is stronger in Germany than in any other EEC country (Papadakis, 1984).

The commitment of the West German government was similar to that of the US NEPA policy in that all projects involving federal authorities required an essential examination of environmental compatibility (Selman, 1981). In 1976 the French government introduced the requirement of EIA for all 'significant' public works and private projects requiring public authorisation. The French approach to EIA is better than that required by the original NEPA legislation in that the majority of private projects which require planning consent must also be subjected to an impact assessment.

In Britain, no formal requirement for EIA exists though an increasing number of major planning decisions now incorporate EIA ideology. For example, a public enquiry for the siting of a new motorway or construction of a major new industry would take into

Table 8.3 Areas of Concern Most Suited to EIA Techniques After O'Riordan and Hey (1976).

Industrial Developments:	Metallurgical projects chemical, textile and leather industries transportation developments such as: motorway alignment, airport expansion new railway construction e.g. TGV in France location of nuclear power stations location of electricity power lines toxic and hazardous waste disposal
Urban Developments:	new town locations urban renewal
Agricultural Developments:	aforestation projects intensive agriculture feeding lots alternative agricultural crops conservation of agricultural landscapes.

account the visual effect such a development would have on the landscape along with an assessment of damage to plant and animal habitats. The impact on the lifestyles of the local human inhabitants would also be included.

Such a study would be based upon somewhat traditional principles, for example a detailed account of the physical geography, the ecology and the socio-economic conditions of the site. Then would follow a study of disturbance during construction, potential hazards, consumption of resources, end products and disposal of waste materials. The study would conclude with consideration of the probable primary and secondary effects which would occur both during construction and in the post-construction phase (Clark et al, 1976).

The European Environmental Bureau (EEB)

It is clear from the previous section that considerable variation exists within the European Community in the level of response to environment and conservation planning. Difficult as it may be to establish national planning strategies which incorporate adequate safeguards for conservation, it is necessary to go one step further and attempt international cooperation for conservation issues.

The restricting of pollution, the safeguarding of species and the establishment of large nature reserves all can involve the crossing of international boundaries. There is a long-established history of dispute between neighbouring countries over transfer of pollutants across international boundaries. (For example, the copper smelter at Trail, British Columbia, released pollutants which allegedly caused damage to forests in the adjacent US state of Washington. More recently sulphur from Britain and West Germany has given rise to acid rain over Scandinavian countries.)

The European Community (EC) established the European Environmental Bureau (EEB) in December 1974 in an attempt to coordinate environmental policy. Its formation owed much to the United Nations Environment Programme established two years previously at the UN Conference on the Human Environment held in Stockholm in 1972.

The EEB represents a coalition of environmental and conservation groups from within the EEC and acts in such a way to further the trans-national interests of conservation and environment in both scientific and planning spheres (Lowe et al, 1983). The EEB is grant aided from EEC funds but yet remains outside the formal EEC structure. The objectives of the EEB are, in many ways, contrary to the official interests of the EEC, that is they are concerned with conservation of resources and not with the development of agricultural and industrial interests.

Impact of the EEB on the EC policy has taken place

very much behind the glare of publicity. For example, in 1980 the EEC announced its first Environmental Action Programme (Ellington & Burke, 1981) in which research money was made available for basic scientific study of environmental and conservation issues.

The Environmental Programme has currently been expanded so that for the period 1986-91 some .40 million has been allocated to two main research areas. These are as follows:

Environmental Protection - £32 million

Aim: to provide a scientific base for the implementation of the EEC's environmental policy and to promote long-term research on important environmental problems in support of a preventive approach to environmental protection.

Specific research areas which will receive funding are identified as:

- ecological effects of pollutants
- assessment of chemicals in the environment
- effect of air quality on terrestrial and aquatic ecosystems
- waste research with emphasis on treatment and recycling
- water quality
- soil quality
- ecosystem research - dynamics and vulnerability
- pollution reduction

Climatology Research - £8 million

The total sum of money allocated for the Environmental Programme is hopelessly inadequate to fund all the research headings in the above list for a period of up to five years. While in no way wishing to denigrate the achievement of the EEB, the emphasis given to conservation and environmental issues at the highest EEC level has been extremely poor. It would be to the everlasting credit of the EEC member countries if they could provide leadership in the harmonious management of environmental resources just as European states in times past provided that stimulus for the Renaissance movement, the Reformation and for the Agricultural and Industrial Revolutions.

Future Role for EIA

The clear objective for environmental impact studies when first introduced during the early 1970s was as a means of clarifying the emotional 'noise' which surrounded many ecological and conservational issues. A more objective and formalised approach to presenting the ecological case has been obtained and EIA is now recognised as an established and relevant planning technique which can stand alongside a whole range of other planning methodologies.

As always, however, the EIA system can be exploited and its critics claim that EIA has become a means whereby objectors to a project can introduce serious delays to the planning process and which can result in economic and social dis-benefit or even the cancellation of projects. Numerous examples of these delaying tactics can be seen in North America and also in Britain - notably in connection with the location of London's third airport, the construction of the Sizewell B nuclear power station and the detailed location of new motorway construction around London.

Lee (1982) has suggested that EIA will become increasingly used in both developed and developing countries and that two main areas of consolidation are possible.

1. A broader view of EIA will be taken with its use extended from specific projects to include wider plans, policies and programmes. In this event, 'tiering' assessments will become necessary. Within a regional plan there may exist several development programmes each comprising a number of projects. Each project will require its own EIA each undertaken not in isolation as at present, but inter-related between projects and also linked vertically to programme and plan.

2. A sifting out of EIA methods. Many of the initial techniques were used because of their simplicity and established use in pre-EIA times. Assessment methods are now required which can:

a) be automated (computerised) so that frequent updates on data are possible

b) are amenable to conversion to algorithms for use in computational analysis

c) are of particular usefulness in their decision making capabilities on proposed action

d) are cost effective in their application thus improving the cost-benefit balance from the use of EIA.

Training the Ecological Planner

The ecological planner is presented with a future in which many complex challenges exist. To be a successful ecological planner will demand the most specialised of training courses. Not only will it be necessary to have studied the biological sciences but also components of physical and human geography. Whether it will be advisable to have a joint degree in Biology and Geography or a degree in the newer subject called Environmental Science will depend on the detailed content of the degree structures. It will certainly be necessary to follow the first degree by a specialised post-graduate course in Planning, in Countryside Management or in Conservation Management.

Such courses already exist and are usually multidisciplinary in nature. A typical 'masters' course in conservation management would include classes in ecology, forestry, agriculture, land economics, computing, geographical information systems as well as the more formal requirements of planning theory and practice.

No amount of formal training can replace the need for the ecological planner to be primarily involved with the practical skills of the discipline. Ecology and conservation issues are involved with real-life components and thus field experience is of utmost relevance. The ecological planner should be exposed to the widest array of geographical examples. International experience is one of the most cost-effective ways of transferring information between countries. Exposure to past attempts at environmental and resource management is another means of acquiring a broadening of experience quickly and cheaply.

Some Other Techniques for Ecological Planning

A wide variety of methods exists for the assessment

of individual value of plants and animals, of ecosystems, habitats and unique sections of the biosphere. Some of the main headings under which individual techniques exist were given in Table 4.2 and reference to some of these has already been made in Chapters Four and Five.

The purpose of this section is to examine the approach most likely to be used by the ecological planner. Many of the techniques described earlier, for example the Biological Yield Assessment and Scarcity Valuation (see pp. 95-98) are specialist techniques and have usually been used in scientific research projects as opposed to practical planning applications.

Planning nowadays requires that a landscape be used in as an effective way as possible. The term 'effective' is taken to mean that a unit of land must be used in such a way that it provides the maximum usefulness commensurate with its inherent biotic and physical capability. For example, an area of flat, well drained land with deep brown earth soils might be well suited for the growth of wheat. It could also successfully grow a range of other crops (oats, grass, trees). Its optimum use would be that which combined the greatest economic yield with the crop which was in greatest need, and the production of which was within the long term sustainability of the site.

Flat, well drained land can also be used for many alternative purposes. Construction costs would be relatively low; the land could be used for housing, for extensive industrial development or for communications. Wherever possible the land areas which are of prime agricultural value should be retained exclusively for that use.

In the poorer land areas (usually the class 3 or 4 land types) the range of ecological options are restricted by physical factors, for example, due to slope, elevation, shallow soils, climatic exposure. These variables inevitably work in complex combinations. They operate in ways which restrict the growth of many plants and animals. As land use becomes even more limited, classes 5 and 6, many species are prevented from competing for living space due to the severity of conditions.

One of the prime objectives of the ecological

planner is the identification of those areas in which ecological options are most seriously limited by physical factors. In the past, this task has been called landscape evaluation and a number of successful techniques have been developed most of which have been devised to suit specific landscapes, landforms or land uses (Crofts et al, 1974). These techniques fall into two categories:

1. Qualitative, in which descriptive classes are used to allocate land to specific groups (Zube et al, 1976). Advantages of this method are speed, cheapness and adaptability. Disadvantages include lack of rigour, lack of agreement between workers, and data which is descriptive in format.

2. Quantitative, in which relationships are established between landscape features and a numerical score. The quality of data can show great variety from the classificatory level e.g. class 1 = good land, class 6 = poor land, to 'interval' and 'ratio' data in which a numerical value has a specific meaning, for example, a pH value of rain-water at 4.5 is more acid than a sample with a value of pH 5.5.

Gilg (1978) has provided a critique of many of the methods used in landscape evaluation. Many of the techniques are based upon the division of landscape into a regular grid square net. Each cell in the net is then assessed according to a classificatory scale.

The use of the grid-square method with which to collect data is called the 'raster method' (Berg, 1985; Jones, 1986). It has many advantages:

1. The size of the grid cell can be altered to suit the intensity of study.

2. Raster-data can be easily and economically handled by computer analysis.

3. Raster-data can easily be edited if new values have to be added.

4. A series of different components can be measured on successive grids and overlays can be drawn in which combinations of variables can be displayed in mapped form.

There are disadvantages to the method, the most significant being:

1. The raster-based map is geometric in appearance and not geographical, the mapped boundaries follow the edges of cells and not true ecological and geographical boundaries.

2. The data base is slow to produce, is consumptive of many man-hours of skilled time and hence is expensive.

The most favoured size of grid cell has been $1km^2$, extensively used by planners in Warwickshire, England (The Councils, 1971) and more recently by the Institute of Terrestrial Ecology in their all-Britain ecological survey (Smith,1983). A grid square of such large size will inevitably contain very considerable internal variation. Davidson and Jones (1986) have used a grid square of 1 hectare in a detailed study of $70km^2$ of the Campsie Fells in Central Scotland. In this study, a grid square of $100m^2$ allowed highly detailed land assessment maps to be produced (see pp. 106-8 and Fig 4.4).

Fernie et al (1985) have presented a thorough review of many techniques appropriate to biosphere assessment. These include:

1. Carrying capacity; defined as a product of management judgement rather than a precisely defined measure - it is a decision-making concept rather than a scientific concept (Hendee et al, 1978). Much work has been undertaken on the carrying capacity of land devoted to agriculture, forestry, recreation, range-land grazing and fishing. It could be usefully extended to include multiple land use assessments in which conservation was one of the land uses.

2. Project assessment; increasingly all land/resource management planning policy is subject to intensive scrutiny. It is no longer sufficient for the ecological planner to produce a series of base maps and a written report. From the 1970s, cost-benefit analysis, use of multiple objectives and discounting rates have been employed, soon followed by environmental impact assessment (see Chapter 4).

3. Cost-benefit studies; this widely used economic technique examines the ratio of estimated benefit values divided by the cost of providing those benefits. The resultant figure provides a unit benefit per unit of cost. Such a figure allows the following questions to be answered:

a. Is there an optimum size for the case study, beyond which dis-economies of scale set it?

b. Which of several, alternative strategies is the most cost-effective?

The disadvantage of cost-benefit studies is its rigid approach to a highly dynamic problem (O'Riordan, 1983). In spite of this rigidity, cost-benefit studies have been widely used, particularly by financial institutions when deciding upon whether to make available loans to third world countries.

One final technique should be mentioned, that of Risk Assessment. Any action involving the biosphere and its components can result in a wide variety of end results (because the biosphere is a probabilistic system, see page 48). Risk assessment methodologies have so far been applied to potentially hazardous installations such as nuclear power stations and chemical works (Fernie et al, 1985).

Similar methodology, based on the probability theory of the most unfavourable situation taking place, could be used to infer the end result of ecological planning. Some ecologically related examples would be:

1. What is the worst effect possible of releasing a known volume of cooling water from a thermal power station into an estuary?

2. What is the worst effect possible on stream run-off if a forested catchment area is clear-felled?

3. What is the worst effect possible of a 25% reduction in precipitation amounts over northern Africa for three consecutive years. How is the situation made worse if rainfall reduction becomes 33%?

Public Participation in Ecological Planning

The task of the ecological planner is to match the ecological capability of an area with the objectives of society (Norton et al, 1982). It is debatable whether the planner should be responsible for initiating a planning objective or whether the planner should respond to public demand. Certainly, the incorporation of an ecological component into the established planning processes during the 1930s-50s owed much to the pressures exerted by small, special interest groups such as the Sierra Club in North America. Nowadays, the action of pressure groups, such as Greenpeace, is still evident. Planners, however, are also more willing to assume responsibility for the initiation and extension of ecological planning.

Unfortunately, the objectives and aspirations of society are rarely defined with clarity. Neither are they static, being determined in part by political and economic considerations. Technological innovation also changes the capacity in which man can interact with the biosphere. Technological advance usually means greater pressure upon the biosphere, but it should also be remembered that greater technological ability also results in a greater capacity to repair a damaged biosphere.

Generally, however, technological innovation is several steps ahead of our ability to recognise biosphere damage. No sooner has the damage been recorded and steps taken to repair it than a new, more advanced form of technology brings about a more serious threat to biosphere stability (see Table 8.4).

Lorrain-Smith (1982) has suggested that it may be necessary for the ecological planner to redirect his approach away from the traditional target of biosphere quality and instead substitute the goal of maximum consumer welfare. Under such an incentive the ecological planner would be concerned with creating environmental conditions which are in accordance with the expectations of the public. But the desired quality of the environment will vary both on a geographical and time basis. Exactly the same economic principles apply to the provision of environmental quality as to a service or availability of goods. Demand level is a function of cost (and the cost of any substitute), the level and distribution of disposable income and of consumer taste.

Willingness to pay for conservation measures through taxation or a participation cost must be assessed against all other goods and services which the public deems to be desirable. Thus, any attempt to provide an enhanced environmental policy becomes part of the overall regional and national planning policy. There is little to be gained by raising the level of conservation planning if the public do not perceive the need for, nor can afford, the expense of such a policy.

The argument outlined by Lorrain-Smith (op cit) must not be applied with total rigidity to conservation issues. The American attitude towards 'wilderness lands', or the British concern with wild animals and ancient rights of way, have become part of the general conservation infrastructure and as such must be considered unique and undefinable inputs to the national conservation policy.

In order to achieve future improvements in the quality of our living standards we must motivate a larger proportion of the general public than the small special interest groups at present. Concern with environmental issues is currently a predominantly middle-class interest (Lowe et al, 1983). A broadening of the social groupings involved in conservation and environmental issues is thus necessary. It is significant that membership levels of 'environmental action groups' show a positive correlation with affluence and development pressure. In Britain, the south east region has 53% of all environmental society members but only 30% of the population. Membership of similar groups throughout the whole of Scotland is less than membership totals for individ-ual counties of Surrey, Kent or Sussex (figures quoted in Lowe et al, 1983).

Increasing affluence, mobility and education standards should, in theory, produce a public who create a greater demand for a high quality environment and at the same time are capable of appreciating the need to set even higher standards for conservation. By participating in the use of the biosphere for recreational purposes (tourism, walking, sailing) it is possible that an increasing proportion of the general public will support the need for more conservation.

TABLE 8.4 Intensification of Air Pollution Hazards and the Technological Solutions Used to Overcome the Hazard Event

Incident	Hazard Event	Management Solution
Particulate Air Pollutants pre-1960	Meuse Valley, Belgium 1930 Donora, Pittsburg 1948 London Smog 1952	Clean Air Acts banning the emission of 'dark smoke'. Effectively reduces particulate output.
Gaseous Air Pollutants 1955 onwards	Technological advances promote high pressure and high temperature combustion (modern automobiles, coal-fired electricity plants, oil fired boilers.)	Fitting of 'lean burn' engines to cars, catalytic converters to exhaust systems, chemical 'scrubbers' to factory chimneys. Taller chimneys releasing pollutants at high temperature and efflux velocity to shoot pollutants higher into atmosphere.
Photochemical Smog 1960 onwards	High-tech industries and mobile society produce more nitrous oxide pollutants thus producing PAN damage.	More expensive cleansing of exhaust, limiting very hazardous processes to specific occasions when atmospheric dilution is at a maximum.
Acid Rain Syndrome 1975 onwards	Gradual acidification of atmosphere and fresh water bodies. High mortality for aquatic animals, conifer trees, mosses, and lichens. Impaired quality of growth in other plants.	Reduction in sulphur levels in industrial air pollutants. Formation of '30% Club' dedicated to an expensive policy to reduce release of sulphur into atmosphere.

Homo sapiens is a remarkable species in so many ways. Our improved capability to study and understand our species, to appreciate the problems created by over-population and the absolute power given by our technological skills has meant that for the first time in our history we have both reason and the methodology whereby we can alter our behaviour so that our chances of long-term survival are improved. It is incumbent upon society to develop the means to adapt to the new circumstances in which it finds itself so that our future in the twenty-first century remains one of expanding environmental possibilities.

Chapter Nine

CONSERVATION, IDEOLOGY AND POLITICS

Conservation of ecosystems and species has become necessary because of the phenomenal increase in the pressures exerted by the human population in its day to day quest for survival. Additionally, the technical capabilities of *Homo sapiens* are such that it is now possible to undertake feats of biospheric 'engineering' of such magnitude and with such rapidity that the time scales operating within the natural world are totally overwhelmed.

Such is the size of the human population ($4.5 x 10^6$) that even if man behaved in the most sympathetic of ways towards the other plants and animals with which we share the biosphere there is little doubt that conflict over resource use would still occur. Man's nature is such that we are not sympathetic towards the biosphere. We mis-treat the physical and biotic resources such that over-exploitation has led to a progressive simplification and impoverishment of the biosphere.

The need for conservation has been recognised for over one hundred years. For example, Marsh (1874) recognised the impending conflict between man and natural resources in North America as did organisations such as the Sierra Club founded in 1892 in the USA and the National Trust (1895) in Britain. As a result of pressure from these sources some limited success was made in conservation.

Conservation Requirement post-1945

The end of the Second World War was marked by a new awareness that mankind was indeed capable of making very significant changes to the physical and biotic resources of the planet. Mineral resources had been used at rates far in excess of pre-war consumption. Maritime blockades had shown how incapable many nations were of sustaining the national economy upon internal resources. International trade in food, fuel and industrial commodities was and remains fundamental to the well-being of the developed world.

The circumstances in which politicians and planners found themselves in the 1950s were such that conservation of ecosystems and species was actively encouraged - see Chapters 5 and 6. Planners of the day took as one of their maxims an earlier definition of conservation, namely that good planning policy would aid resource use and ensure the greatest good for the greatest number for the longest time. By the late 1950s conservation of resources was a recognisable and accepted requirement at both local and national level.

At about this time (1950s) a split became apparent in the motivation for conservation. Cotgrove (1976) has described the two schools of conservation as:

i) traditional conservation, and

ii) liberal conservation attitudes.

Both Tivy (1977) and Sandbach (1980) have preferred to apply the term scientific or ecological conservation to the traditional conservation attitude.

The traditional role of conservation recognised the need to sustain a diverse and stable physical and biological environment. In its purest form traditional conservation policy would insist that the advantages of human technological or economic change must be assessed on the basis of their impact on the physical/biological environment. This type of conservation is an elitist approach favoured by the pure scientist. The methods and arguments used to support its arguments are often complex, time consuming to develop and, apart from the eastern bloc countries, tend not to find favour with politicians (see pp. 143-144).

The second approach to conservation, that of the liberal conservation attitude is founded more upon ethical and aesthetic reasons and places less emphasis upon scientific argument. Sandbach (op cit) has summarised the objectives of this lobby: "Objectivity and rationality impose a distance between the scientist and the subject of investigation. This, together with a depersonalized scientific jargon, allows the legitimate abuse of nature under the pretext of furthering knowledge, of achieving technological progress, or of contributing to economic growth". Such an approach when taken to its extreme development, has given rise to an 'Alternative Technology' movement in which low-technology solutions are advocated in place of the 'high-tech' development which characterises most modern industrialised societies.

Although the alternative technology approach to biosphere mis-use still attracts considerable support, Green (1985) has suggested that a more pragmatic approach can now be discerned in which the arguments of utilitarianism and ethical factors are used to support and advance the conservation cause.

Laudable though the objectives of both ecological camps may be there is evidence which suggests that a wide discrepancy exists between the level of conservation deemed necessary by ecological scientists and the actual achievements of conservation planners (Abdalla, 1977). While governments pay lip service to the needs of conservation of the biosphere, few can substantiate the rhetoric with action (Bartlemus, 1986).

One of the few instances of a national government taking major action on environmental/conservation grounds can be found in the approach of the Swedish government towards electrical energy generation. Sweden has followed a conventional approach to provision of energy needs. In the early 1970s a policy was implemented whereby sufficient nuclear-powered generating stations were planned to supply approximately 50% of electrical energy needs by 2000 AD.

However, by 1980 public concern in Sweden over disposal of nuclear waste from power stations along with fears of industrial accidents and terrorist attacks on nuclear power stations was sufficient to persuade the government to de-commission <u>all</u> nuclear power stations and to make Sweden a nuclear free country by the year 2000. This decision has been implemented in the full knowledge that Sweden does not possess any significant reserves of fossil fuel.

Energy demands post-2000AD in Sweden will be met from renewable resources (HEP, wind and wave power) and from thermal-powered generating stations using heat from the combustion of refuse. Major energy savings are planned from the improved design and insulation of houses (already of a very high standard), an extension of district heating schemes and from improvements in the efficiency of generation and transmission of power.

The elimination of nuclear-powered generating stations from Sweden is an excellent example of how public attitude towards environmental issues has changed over a period of approximately twenty years. During the 1960s public interest and involvement in conservation was confined to very small sectors of society - mainly the young, professional class, whereas by the 1980s the situation, especially in Scandinavian countries, had extended to become a major concern with most sectors of the general public.

No matter how strong the scientific, utilitarian and ethical arguments may be, nor how extensive or well enforced the legislation connected with conservation, it is the willing cooperation of the individual citizen which determines the success or failure of a conservation policy.

The conservation ethic belongs to a small number of uniquely humanistic behavioural traits. Along with such features as religion, political organisation and the creation of a welfare programme to support the deprived members of our society, the act of conservation is peculiar only to the human animal.

Conservation differs, however, from the other uniquely human characteristics in that, if successful, a conservation programme should eliminate the reasons for its existence! Dasmann (1972) recognised this condition when he commented that a conservation policy could be judged a success when it was no longer required.

Even the most conservation-conscious countries such as Sweden are some way from eliminating the need for conservation. Most countries are still making the first tentative steps along the conservation pathway. Hibbard (1972) has shown that it takes approximately 25 years for a concept or a process to pass from the development stage through to general acceptance and full implementation. This is the so-called delay time, or 'lag'. The lag time can exceed 25 years when the concept is revolutionary in nature.

Lag time is an important factor in conservation issues, none more so than when it is realised that the attitudes currently applied to conservation schemes will still be working through the system in the early decades of the next century.

It is vital, therefore, that we continue to develop an advanced strategy for conservation and that research and experimentation into improved conservation techniques receive sufficient priority so that we can meet the conservation requirements of the twenty-first century.

A major achievement for the conservation movement would be the persuasion of the engineer, industrial designer and economist to incorporate a 'conservation input' into all industrial projects currently at drawing board stage.This has been one of the aims of NEPA legislation (see pp. 205-207). Incorporation of similar control measures in all new development work would make a significant advance in conservation achievement.

At present our conservation structures are 'add-on' items, akin to the optional extra purchased with a new car. The optional extra approach is expensive; if the component is designed into the structure from the outset then significant cost savings can be achieved.

An example of the optional extra principle at work in conservation issues can be seen to operate in the generation of electricity in conventional coal-fired power stations. In Britain coal-fired power stations are based upon specifications established in the 1960s. Flue gas cleaning devices are confined to precipitators which collect the particulate matter, especially the fly-ash material, the finest material of about five microns in size. Little attempt is made to remove the gaseous pollutants in the flue gases.

The main gas to escape from a coal-fired power station is sulphur dioxide. This can be almost totally removed from the flue gas but of course increases the operational cost of the generating stations. This increase, if met fully by the consumer, would add an estimated 5% to the cost of electricity. In Britain, where electricity is already an expensive commodity compared to its cost in many other countries, notably New Zealand with its abundant hydro-power, the additional 5% has been judged to be unacceptably high and as such flue gases are allowed to escape into the atmosphere fully laden with SO_2, H_2F and many other gases in lesser quantities. To retro-fit gas cleaning apparatus to twelve large coal-fired stations in Britain would cost £2,000 million (Government press release, 1986).

Conservation ideology has made major advances in its first one hundred and twenty years of operation. During this time the first national park was established (see p.115) and was followed by other parks, reserves, sanctuaries and sites of special scientific interest in almost every other country in the world. Numerous other conservation 'keystones' have been established, for example the concept of the 'land ethic', (see p.94) (Leopold, 1949) and the World Conservation Ethic (p.199) (IUCN, 1980).

The relevance of conservation is still not wholly accepted by those who are responsible for governing and shaping our societies. The predominance of economic and political considerations are two major impediments to the universal acceptance of the principles of conservation and it is necessary to look briefly at these two areas of resistance (United Nations, 1984).

Conservation and Economics

A fundamental difference exists between an economic definition of a biospheric resource and that of an ecological definition. To the economist, a resource is judged entirely upon its financial value; there is no consideration of its nature or inherent characteristics (Coddington, 1974). Thus the value of a resource is determined by the demand for that resource and also the technical difficulties in making available the resource. A resource which is difficult to develop, extract and process will be more expensive than a widely available resource (Beckerman, 1974). Thus North Sea oil is an expensive resource whereas water as a requisite for industrial processes is usually a very cheap commodity (Cottrell, 1978).

The economist would consider the conservation need in two different ways. In both instances conservation provision is assessed in financial terms; there is little or no involvement in assessing the biological or ecological costs of conservation.

External Effects

The first area for consideration is the so-called 'spillover effect' or externality factor. For example, a consequence of most industrial production processes and also of many consumer processes is the creation of unintentional side-effects on the environment. Common side-effects would be pollution of air, water and land or the removal of a specific habitat as a result of mineral extraction. Externalities are notoriously difficult to predict; once identified they also prove very difficult to cost. As a result, external factors tend, wherever possible, to be ignored. This deliberate avoidance of the effect and cost of the economic system upon the ecological system is of fundamental importance in explaining the attitude which man has traditionally displayed towards the biosphere.

The fault does not lie entirely at the door of the economist. If the ecologist or conservation manager could calculate the financial damage caused by an economic process on an ecosystem, then that value could be translated into an economic input and could be entered as a debt against production costs. In this way the cost to the environment would have been changed from an unknown externality to a known internal cost (Freeman et al, 1973).

While accepting the difficulties in costing the value of a conservation externality it should be possible for governments to insist upon one of two strategies to finance the cost of conservation.

1. The setting up of a 'conservation fund' financed by a levy or taxation on industrial profits. The fund would be administered jointly by the government departments of finance and of environment and would finance research into, and the running costs of, conservation projects.

2. A compulsory contingency fund comprising a fixed percentage figure of annual company turnover to be used to finance government-

directed conservation projects.

Either method of supporting conservation projects would create a very real responsibility for the conservation manager and planner. It would be necessary to calculate the effect of economic activities on the biosphere and its resources. The contingency fund would then be used to redress the human impact on the biosphere. A financial balance sheet could be produced to show the industrialist, the economist, the politician and the public how the fund was spent and its achievement in terms of conservation gain.

The use of monetary values to measure the success of conservation is disliked by the conservation purist. An elitist attitude demands that an ecological resource be judged on its scientific or aesthetic value. Such an approach is not helpful in the immediate short term. The economist must be helped in his task of internalising the cost of economic activities on ecosystems and species. If a financial argument is the only one the economist can understand then ecologists must learn to use this approach.

If we assume that only about 25-33% of environmental damage can be internalised then considerable environmental damage still remains for which no compensation is received. There are some other ways by which recompense can be gained.

The most obvious is through imposition of an 'environmental penalty payment' which would be levied on the most environmentally damaging industries, for example chemical factories, oil-from-coal extraction plants, dyestuffs, pesticides and ceramic factories. Penalty payment levels would be in addition to the alternative methods of financing basic conservation measures explained above. Penalty payments could be avoided only if the company could show that an attempt, in financial terms equal to the penalty payment, had been made to reduce the degradation effect upon the biosphere.

There are few industrial processes for which no additional cleansing devices exist. If industrial designers knew that penal taxation measures would be applied to any process which destroyed environmental resources then the stimulus would be given to the design of more efficient equipment.

It would be also possible for some human activities to enhance the environment. New towns with segregated urban and industrial areas could be planned with wildlife corridors to connect undisturbed habitats. Incentive grants for positive contributions to conservation should be payable from government funds.

Through a combination of taxation and incentive payments it should be possible to raise the profile of conservation achievements. One major problem remains: who will assume responsibility for setting conservation standards? who will monitor conservation achievements? how will non-compliant industries be encouraged to participate in active conservation issues?

One is forced to conclude that a taxation subsidy and incentive approach to ecosystem and conservation management would be complex, contentious and expensive to implement.

As an alternative to taxation and subsidies, Coddington (1974) has suggested that a somewhat simpler system might be that of 'compensation' in which the disadvantaged party was paid a financial sum for loss of conservation value. This method suffers the problem of who shall arbitrate on the issues of quality decline and also of how much compensation will be paid.

Compensation might be a once-off payment. For example, the construction of a new motorway or extension of an airport runway could justify the payment of a grant for soundproofing expenses. Compensation payments are best suited for human disturbance and can deal less adequately with natural ecosystem disruption.

A common problem to all these methods is the attempt to apply contemporary conservation attitudes to an economic infrastructure which may be based upon industrial, economic and planning policies which are between 50 and 150 years old. What was environmentally acceptable to previous generations is now unacceptable; equally true, the conservation standards of today will become rapidly outdated in the future. The most pragmatic view would be to insist upon the highest design standards for any new man-made development.

Degradation of Biosphere Resources

One conservation issue does command the interest of the economist: that of availability of resources for economic use.

The economist is involved only with those biospheric resources which are currently used by man in industrial and commercial activities. Thus, a resource can be water - usually fresh water for cooling or dilution purposes. Other biosphere resources include soil, trees, agricultural crops, minerals, stones and aggregates, and not forgetting land itself. Land is the ultimate resource. It is the substratum on which man passes the majority of his life (see Chapter Two) and as such is a financially valuable commodity.

The division of biospheric resources into renewable and non-renewable parts (see pp. 7-8) is as much an economic division as it is an ecological division. The economist is less interested in the mechanisms of renewal of a resource and more in the costs involved in exploiting the resource. As the resource becomes scarcer, and assuming demand remains constant, then so the value of the resource will increase. It may become so expensive that it exceeds the consumer's willingness or capability to pay. If so, the economist will resort to:

1. Recycling of used materials, already widely practised for many metals (iron and steel, copper, lead, zinc); paper; glass; timber; fibre (wool, cotton) and plastics. The economics of recycling has been a relatively recent phenomenon and has necessitated totally new industrial techniques.

2. Increasing the life span of components. Improved design of components can often extend the life of a component or, if the component fails, good design can help facilitate its repair. Conversely, a component can be designed to be replaced after a finite time. Industrial and domestic equipment has benefited from the inclusion of self-checking diagnostic circuits controlled by micro-chips. The trends towards prolonging the life of a commodity is in contrast to the planned obsolescence which was highlighted so vividly by Packard (1960). Public preference for longer lasting and repairable

components is reflected in a willingness to pay higher prices and which can lead to higher retailing profits.

3. Replacement of non-renewable resources by renewable, in particular, renewable energy resources (solar power, tidal power, HEP, biomass) are seen as replacements for the increasingly expensive fossil fuels and the catastrophe-prone nuclear fuels. Increasing consumer affluence permits payment of a premium price for resources with a perceived high quality produced at minimal environmental cost. Unfortunately, little if any of the income from these commodities finds its way back into practical conservation issues.

Economic Versus Ecological Time Scales

Ecological time scales extend over many hundreds of years. The economist cannot afford the luxury of working over similar time spans and instead is accustomed to working on five or ten year development plans dictated by our political systems.

Processes which may be acceptable over a short economic term become hopelessly inefficient when assessed over a longer time span, for example the replacement of modern steam locomotives by new diesel locomotives in the 1950s and '60s. In this example a reduction in immediate operating costs outweighed the comparative future fuel costs.

The economist must be persuaded that it is more profitable to move away from exploitive short-lasting systems to longer-lasting systems which show steady-state (homeostatic) characteristics (Daly, 1977). Such a change is fraught with problems, not least the generally held view that a move from an economic system based upon growth and development to one of steady-state would involve a decline in material wealth leading to recession and unemployment.

The 1980s have witnessed spiralling unemployment levels in all the industrialised nations apart from Japan. This has occurred because of an international trade recession and <u>not</u> because of a change in the economic system from a growth-orientated system to a steady-state system. The Keynsian economic principle of growth may no longer provide the ideal model for western nations to follow.

In an effort to improve productivity, automation has replaced jobs (Kahn, 1976). Very little attention has been given by economists to the hidden costs of job redundancy. Additionally, robotised lines are constructed from metals, many of which are scarce and expensive, and consume electrical power and lubricant oils. The use of robots in dangerous working environments (paint spraying, or inside chambers with high levels of toxic gases) is highly desirable but their use to boost productivity in non-hazardous situations may not be cost effective when measured in wider terms.

Ecodevelopment and Employment

A commonly held belief is that the pursuit of ecologically orientated goals will result in fewer jobs than if economic goals are set. Cotgrove (1982) has provided statistical evidence on this issue (see p. 236). This attitude reflects the prevalant view amongst many sectors of the public that ecologic and conservation goals are anti-growth and anti-development. This view is incorrect. Incorporation of EIA evaluators into the economic and industrial environments would encourage both a growth and diversification of employment prospects.

The type of employment would change. There would be a move away from repetitive 'production-line' jobs to those involving monitoring the constantly changing ecological conditions. Field surveyors, experimental officers, statisticians, technicians, instrument makers, conservation planners and economists all would be reqired. A totally new employment stratum would develop, namely the environmental assessment group.

For such a change to come about would require much more than a replacement of economic values by ecodevelopment values. It would need a major change in the social and political perspectives of western man (Theobald, 1970). In place of indicators such as GNP we would require a Gross Stability Factor or a Gross National Quality - an index of constancy based upon a series of measurements, some economic, some social and some ecological indicators.

Future historians will, no doubt, ponder upon the social and political circumstances which operated between the years 1935 and 1999 and which produced a civilisation obsessed with just one criterion of success - that of growth. Much of that attitude has been due to the philosophy of one British economist, John M. Keynes (1883-1946), who argued that governments should direct themselves towards providing a legal and fiscal environment which would support maximum production, maximum consumption and minimum unemployment (Hansen, 1953).

Keynsian economic theory pays little regard to the ability of our biosphere to support sustained growth. As such it is an outdated and inappropriate theory on which to base future development of society. What is required is a new model which can provide international stability but incurring costs which can be sustained by our biosphere.

Sandbach (1980) has provided a coherent review of some possible strategies:

1. The Keynsian model can be followed until the exponential growth curve crashes.

2. An alternative, radical model is imposed upon an industrialised society disillusioned by the contemporary situation in which the discrepancy between the developed nations has risen to alarming proportions.

3. A new economic and political strategy can be developed in which steady-state replaces the growth conditions which currently operate.

Conservation and Politics

For the most part, conservation issues still do not play a major influencing role in political decision making. Papadakis (1984) has presented evidence to show that at best only 10% of voters supported conservation and environmental matters at times of national elections. Interest and concern in conservation issues on behalf of the general public show extreme fluctuation. There is little doubt that the late 1960s and early 1970s saw maximum media publicity given to such issues as population pressure, pollution problems, food supply and role of the oceans as future living areas. Morton-Boyd (1972) considered that environmental concern was already declining in 1972 while Bowman (1975) recognised further decline three years later.

Major catastrophes such as the escape of toxic gas in 1985 at Bophal in India or the Chernobyl nuclear power station accident in May 1986 produced an upsurge of public concern which was reflected in a temporary recognition that political attention to conservation and ecological policy was sadly inferior to the attention politicians pay to many other issues such as trade and defence and the arms race.

Unless substantial changes occur in the ways in which we treat our biosphere then it is now probable that a serious problem will occur before the end of the current century. This catastrophe may be a serious famine, or nuclear accident from a civil or a military source or it could be warfare brought about by the inequalities which exist between the developed northern hemisphere countries and the poorer, mainly southern hemisphere, nations. Miller and Armstrong (1982) have shown that despite the strength of the scientific and ecological case for change, and even the economists' acceptance that some change in the economic growth model will probably be necessary in the near future, there is still little evidence that politicians will allow the necessary change to occur. Unwillingness amongst politicians to initiate change is not unusual unless the politicians are radicals who seek to gain power!

Governments throughout the world have shown very varied response to the need to foster and support public and political changes towards conservation issues. The degree of action is usually positively related to the strength of the 'pressure group' lobby. In this respect, government response is no different from that which marked the very beginning of the conservation movement in the 1860 (see pp. 2-3).

The action of pressure groups has taken on a new dimension over the last twenty years. A significant step was taken in 1972 when 38 eminent British scientists signed a document entitled A Blueprint for Survival (The Ecologist, 1972). In it a direct attack was made upon the prevailing economic and political systems which had encouraged the exploitation of the biosphere and its resources. From that statement was created a new grouping, The Ecology Party, which had as one of its aims the persuasion of established political parties to include a conservation policy within its manifesto.

The response by the established political parties was directly opposed to the ecological argument. Typical of the response was that of the late Anthony Crosland, the Labour Secretary of State for Environment (1974-76) who stated: 'To say that we must attend meticulously to the environmental case does not mean that we must go to the other extreme and wholly neglect the economic case. Their approach (of the conservationist lobby) is hostile to growth and indifferent to the needs of ordinary people. It has a manifest class bias and reflects a set of middle and upper-class value judgements for which preservation of the status quo is the sole consideration' (Crosland, 1971).

The three criticisms contained in this statement, anti-economic approach, anti-growth and support of the status quo, suggest that the politicians' understanding as displayed by Crosland, was hopelessly out of touch with the real aims of the conservation movement. The most blatant mistake was the belief that the conservationist supported the status quo.

A considerable volume of research now exists in which the motives and the support base for conservation have been evaluated (Lowe and Goyder, 1983). Some 60% of the population involved in a survey indicated a very considerable concern with issues such as pollution, famine and over population. Cotgrove (1982) found a majority (64%) in favour of imposing a pollution tax on major producers of enviornmental pollution.

If a majority of the electorate have indicated a real concern over environment and conservation issues, why is it that more rapid political progress has not been made towards incorporating such items into electoral manifestos? It is easy to be cynical and suggest that if any of the major parties were to include a significant conservation policy in their manifesto then their financial backers (industry or trade unions) might withdraw their financial contribution in fear that an increased conservation awareness would bring discredit or reduce the market potential or reduce employment possibilities for industry and commerce.

Kennet (1974) maintains that no politician who has gained credibility would make a sudden, rash decision either for or against the conservation lobby. In order to retain power, the politician seeks to reconcile the interests of all the electorate within his constituency. Such a behavioural trait has both advantages and disadvantages. It prevents wild swings of political extremism and thus helps maintain society on a steady course. Alternatively, it hinders acceptance of any bright new idea and preserves the political system as a predominantly conservative body.

Apart from West Germany and the Scandinavian countries conservation issues have not yet become major national election issues. As such, conservation policy is generally considered to be a secondary conflict area (see Table 9.1). In almost every case, politicians will be unfamiliar with the detail of conservation arguments. A research assistant will be employed to prepare a report on a specific conservation issue which may be relevant to the constituency. If the politician is a Cabinet Minister then advice can be provided from the Scientific Civil Service. But if the politician is a seasoned campaigner then, as Kennet (op cit) asserts, views will be taken of local environmental pressure groups. Armed with the necessary information the politician can then speak and vote on the issue with all necessary confidence.

Table 9.1 Primary and Secondary Conflict Areas in Politics

Primary conflict	Political parties are diametrically opposed to each other in terms of policy. The electorate have no difficulty in differentiating between party policy.
Secondary conflict	General acceptance between the parties about policy and agreement that action must be taken. Disagreement about the detailed mechanisms and rate of action. Electorate often confused between broadly similar policies.

Complete reliance on a political solution for conservation issues is not practical. The law of the land relating to conservation and environmental issues cannot determine in detail what conservation action must be followed in every single situation. All it can achieve is who shall decide what can be done (usually a minister) and what general rules should apply to any decision making process.

'In Trust' Resources

Conservationists have for some time argued that the physical resources of the biosphere (the land, seas and air) should be held in a form of 'national trust' for the perpetual benefit of the public. The well being of the trust properties should be the responsibility of local, county or regional and national government. Ideally, it should be possible to incorporate the necessary legal framework to safeguard trust land into the national and/or regional legislative framework. Only Czechoslovakia has made any significant progress in this direction. It is significant that this country also has some of the best organised national parks.

The USA also has a series of important legislative measures aimed at safeguarding the needs of conservation. In particular, the National Environmental Policy Act, 1970 has been described (see pp. 205-207) as one of the most important pieces of environmental legislation ever passed.

The American legislation has been copied by several other countries, notably Australia and New Zealand and some Canadian states. In Britain an alternative strategy has been adopted, that of the Public Enquiry. The Town and Country Planning Act of 1971 stipulated that any major land use change must be preceded by public enquiry (Smith, 1975). Any interested and informed person can present evidence at the enquiry and a decision is reached by a presiding inspector. The decision is usually, though by no means always, accepted by the minister for the Department of the Environment. The public enquiry is a genuine attempt to involve the public in determining what should happen to environmental resources. In recent years the procedure has been abused by both government and local pressure groups through delaying tactics and an unwillingness to accept the findings of the enquiry.

Conclusion

It is government which is responsible for taking the necessary action to protect the individual citizen and the environment and its resources. By tradition governments have responded slowly to pressure of all forms; this is certainly true for environmental issues. As this chapter was being prepared the British Government announced a £600 million project to remove sulphur from the exhaust plumes of three coal-fired power stations. It has taken exactly 30 years for government to remedy a deficiency contained in the original British Clean Air Act of 1956. Conservationists have pressurised the government for legislation to control gaseous pollutants. International pressure (EEC and Scandinavian countries) has eventually forced the British government into limited action. Thirty years of delay has caused untold environmental damage in the form of species and habitat loss, much of which can never be redressed.

As man moves forward to the next century it becomes imperative that our social, ethical and legal responses to the biosphere and to all its inhabitants become faster, more sympathetic and more multinational in approach. These demands represent major challenges to the maturity of our societies. No longer can we pursue local or even national solutions to biosphere problems.

Conservation of ecosystems and species requires an international response. We have developed two of the necessary levels of operation: we live our daily lives at a local level, concerned with local issues at family or community level; our historical development allows us to also appreciate the need for national unity, bound by nation laws, politics and economic structures. What we must now develop is a global loyalty in which the security of our planetary resources is recognised as being of paramount importance to the successful operation of the local and national structures.

APPENDIX A Plant Species with Medicinal Uses from Allen (1980)

Plant Source	Drug obtained	Treatment
Glorisa superba	colchicine	anti-inflammatory, anti-gout
Cinchona	quinine	malaria
Catharanthus	vincristine vinblastine raubasine	anti-cancer anti-hypertension
Macuna prurieus	l-Dopa	Parkinson's disease
Rauvolfia	reserpine	cardio-vascular disease
Vinca minor	vincamine	
Ammi majus	xanthotoxin	skin infections
Centella asiatica	Asiaticoside	
Duboisia myoporoides		anti-spasmodics
Hyoxyamus		
Glycyrrhiza glabra		ulcer treatment
Pilocarpus	Pilocarpine	eye treatments
Clibadium sylvestre		heart surgery
Pariera plant	tubocurarine	muscle relaxant
May apple	Ve Pesid	testicular cancer

APPENDIX B Animal Extracts with Medicinal Uses from Allen (1980)

Animal Extract	Treatment
Snake venom	Non-addictive pain killer
Bee venom	Arthritis treatment
Alanotin from blowfly larvae	Assists deep wounds to heal
Cantharidin from European Blister beetle	Treatment of uro-genital complaints
Ara-A from *Tethya crypta* (Caribbean sponge)	Herpes encephalitis
Ara-C from *Tethya crypta*	Leukaemia treatment
Squaline from shark liver oil	Bactericide

APPENDIX C Animal Species on the Verge of Extinction

Case Study 1. Californian Condor.

Phylum: Vertebrata
Class: Aves
Order: Falconiformes
Family: Cathartidae
Genus & Species: *Gymnogyps californianus.*

This species is the largest American land bird and as such has attracted the attention of hunters and collectors for about 100 years. It is a protected species. Unfortunately the female lays but one egg in each breeding season and the developing chick remains in the nest for a period of five months. Both these circumstances enhance the vulnerability of the species. A further reproductive disadvantage is that mating occurs only in alternate years. It is unlikely that the Californian Condor was ever a numerous species. Its territory is large, from edge to edge the territory may extend to 100 km. The preferred habitat is one of remote mountainous areas between 500 and 2000 metres above sea level in which open mixed forest gives way to alpine grassland. Carnivorous in feeding habit, the species feeds mainly on already dead animals; only rarely will it feed on freshly killed meat. In spite of the low reproductive capacity of this bird, it has survived because its habitat has been distant and unspoilt by man, and it lives to considerable old age - 40 years being possible. Since the late 1950s an extension of commercial logging to timberline locations, an upsurge in sophisticated hunting techniques - high powered rifles, fourwheel drive vehicles to gain access to remote sites, and even helicopter transportation into the most inaccessible sites, has pushed the Californian Condor to the edge of extinction.

Year	Total Population of Californian Condors
1950	60
1963	42 (including 20 immature adults)
1966	51
1969	40
1985	21 (5 wild, 1 male, 4 female, 16 in reserves)

In the mid 1970s it became evident that this species would soon become extinct. Poisoned carrion added to the list of problems to be faced by the condor. Most of the Californian Condor were captured and a controlled breeding experiment begun. It has not succeeded. Some conservationists now claim the captive condors should be released so that their final years on this planet can be spent in freedom.

Case Study 2. Black Footed Ferret.

Phylum: Vertebrata
Class: Mammalalia
Order: Carnivora
Family: Mustilidae
Genus & Species: Putonis (Mustela) furo

This species is confined to the state of Wyoming, U.S.A. It inhabits open, forested country and feeds almost exclusively on the Prairie dog (*Cynomys ludovicianus*). The species was believed to have been extinct by 1960 but was rediscovered in 1981. A small, but vigorous population developed only to slump to 58 members in 1984. By early 1986 it was thought that only one existed in the wild with six being held in captivity (4 male, 2 female). This species seems certain to become extinct. The reason for the sudden decline in numbers in 1984 was due in part to a rapid decrease in the population of Prairie dogs which had contracted bubonic plague. This fact alone was not considered to be responsible for the decline in the Black Footed Ferret. It now appears that an outbreak of canine distemper took a large toll on ferret numbers and the breeding population could not develop a natural resistance to the disease.

The status of the ferret in the Mustelidae family is a controversial one. There are no fossil remains of ferrets. It appears the species is a deliberate domestication on the part of man, possibly being bred as early as 1000 BC as a hunting animal. More ferocious than any small hunting dog, the ferret will flush out other mammals from subterranean burrows. It is probable that the Black Footed Ferret is a feral species in Wyoming, and has possibly inter-bred with other members of the _Mustelidae_ family.

Should this species - a product of our own civilisation - be afforded conservation status?

Case Study 3 Sturgeon

Phylum: _Vertebrata_
Class: _Osteichthytes_
Order: _Acipenserformes_
Family: _Acipenseridae_
Genus & Species: _Acipenser ruthensus_

The sturgeon is one of a small group of fish which is born in the upper reaches of a fresh water river, spends its adulthood in the ocean but returns to freshwater each year to breed. In this respect it is similar to the salmon. One other characteristic links both these species - the commercial value of the fish. The sturgeon is the source of that exotic and highly expensive food - caviar, the unlaid eggs (roes) of the species.

Overfishing by ruthless commercial companies combined with the degradation of the quality of water by pollutants in the breeding reaches has reduced this species to the verge of extinction. Its migratory lifestyle has meant that it is impossible to subject this fish to modern fish farming methods. Besides the valuable roe, the meat commands good prices, sturgeon oil is used in medicines while the air bladder provides a substance called isinglass, a product used in industrial processing.

The perilous state of the sturgeon population has been brought about mainly by over fishing. This is a classic example of a food industry exploiting a natural resource to the point of extinction; by so doing it also destroys the food industry itself. It should be possible to apply a conservation management policy based upon the life history of the sturgeon and habitat availability so that a balanced harvest of sturgeon will be available to support a sustainable yield of sturgeon products.

Case Study 4. Falkland Islands Dog.

Phylum: Vertebrata
Class: Mammalia
Order: Carnivora
Family: Canidae
Genus & Species: *Dusicyon australis*

Remote islands are often the location of unique species e.g. Galapagos Islands & New Zealand fauna. The Falkland Islands were sighted in 1592 but there is no record of a human visit until 1690 when a Captain John Strong made land. He recorded the presence of wild dogs on the island; one of the dogs was captured, taken on board but after several months was lost over-board. It was not until 1766 that semi-permanent settlement began. The first inhabitants found large packs of inquisitive dogs which, like so many island animals, showed no fear of the newly arrived humans. The sailors repaid the curiosity of the dogs by killing them simply by setting fire to the tall grass.

The Falkland Islands dog lived in burrows excavated into the dunes. In both appearance and behaviour the animal had affinities with the fox. Its origin is dubious but it might have been a survivor from a shipwreck, or may have been a migrant which was transported on a floating 'island' of debris washed from the South American continent. It survived in the Falkland Islands by eating penguins, seals and geese. Charles Darwin studied the creature in 1839 and predicted its rapid extinction at the hand of man. The only evidence of the creature is Darwin's report along with a dozen or so skulls and some carefully preserved skins in the museums of Paris and London.

BIBLIOGRAPHY

Abdalla, I.S. (1977) 'Development Planning Reconsidered' in J.J. Nossin (ed.), Surveys for Development, A Multidisciplinary Approach. Elsevier, Amsterdam, pp. 151-167

Acocks, J.P.H. (1975) Veld Types of South Africa, Memoirs of the Botanical Survey of South Africa, No 40. Botanical Research Institute, Pretoria

Ahmed, Y.J. and Sammy, G.K. (1985) Guide Lines to Environmental Impact Assessment in Developing Countries, Hodder and Stoughton, London

Ajayi, S.S. (1979) Utilisation of Forest Wildlife in West Africa, F.A.O., Rome

Allen, P. (1814) History of the Expedition Under the Command of Captains Lewis and Clark. Bradford & Inskeep, New York. Republished 1966 as March of America Fascimile Series, No 56, Ann Arbor, Univ. Microfilms Inc.

Allen, R. (1980) How to Save the World, Kogan Page Ltd.

Allen, S.W. and Leonard, J.W. (1966) Conserving Natural Resources, McGraw Hill, New York

Andrewartha, H.G. & Birch, L.C. (1974) The Distribution and Abundance of Animals: A Pioneering Study of Animal Ecology. Univ. Chicago Press, Chicago

Attenborough, D. (1979) Life on Earth, Readers Digest Association Ltd., London

Baker, H.G. (1970) Plants and Civilisation, 2nd. ed., Macmillan Press Ltd., London

Barber, J.T. (1803) A Tour Through South Wales and Monmouthshire, J. Nichols and Son

Barbour, I.G. ed. (1973) Western Man and Environmental Ethics, Addison-Wesley, Reading, Mass.

Barnard, C. ed. (1966) Grasses and Grassland, Macmillan, London

Bartlemus, P. (1986) Environment and Development, Allen and Unwin, London

Bayliss-Smith, T.P. (1982) The Ecology of Agricultural Systems, Cambridge University Press, Cambridge

Bazilivich, N.I.; Rodin, L.Y. & Rozov, N.N. (1971) 'Geographical Aspects of Productivity', Soviet Geography, 12, 293-317

Beazley, R. (1967) 'Conservation-Decision Making; a Rationalization'. Natural Resources Journal, 7, 345-360

Beckerman, W. (1974) In Defence of Economic Growth. Cape, London

Beer, S. (1967) Cybernetics and Management. English Universities Press Ltd., London

Bell, M. (1975) Britain's National Parks, David & Charles, Newton Abbott

Beresford, T. (1975) We Plough the Fields, Penguin Books, Harmondsworth

Berg, A. v.d. (1985) The Use of the MAP2 Program in Landscape Planning and Research. Research Inst. for For. & Landscape Planning. De Dorschkamp, Wageningen, The Netherlands

Bibbey, J.S. & Macney, D. (1977) Land Use Capability Classification, Technological Monograph No 1, The Soil Survey, Rothamstead, Herts

Birch, J.W. (1973) 'Geography and Resource Management', J.Envir.Manag., 1, 3-11

Bliss, L.C. (1971) Conservation, Man and Manipulation Within the Tundras, IBP Tundra Biome. Proc. IV Int.Meet. Biol. Prod. 111-112

Boardman, R. (1981) International Organisation and the Conservation of Nature, Macmillan, London

Bookchin, M. (1974) Our Synthetic Environment, Harper Colophon, New York

Borman, F.H. and Likens, G.E. (1971) 'The Nutrient Cycles of an Ecosystem', Scientific American, 223 pp. 92-101

Boulding, K. (1966) 'The Economics of the Coming Spaceship Earth' pp. 3-14 in Environmental Quality in a Growing Economy. Resources for the Future. Johns Hopkins Uni.Press, Baltimore

Bourne, R. (1978) Assault on the Amazon, Gollancz, London

Bowman, J.S. (1975) 'The Ecology Movement', J. Environmental Studies, 8 pp. 91-97

Brain, C.K. (1967) 'The Transvaal Museum's Fossil Project at Swartkrans' S.Afr.J.Sci., 63, 378-386

Brandt, W. (1980) North-South: A Programme for Survival, Report of the Independent Commission on International Development Issues. Pan Books, London

Breeden, S. and Breeden, K. (1975) 'Dangerous Days for Eucalypt Forests', Our Magnificent Wildlife, pp. 146-149, Readers Digest Books, New York

Bridges, E.M. (1986) 'The Outstanding Gower', The Geographical Magazine, Vol. LVIII, pp.236-239

British Museum, (1979) Dinosaurs and their Living Relatives, British Museum (Natural History) C.U.P. London

Bronowski, J. (1973) The Ascent of Man, BBC, London

Brown, T. (1986) 'A Life on a Blasted Land', The Geog. Mag., LVIII(10), 518-523

Buckman, H.O. & Brady, N.C. (1970) The Nature and Properties of Soils, 7th. ed. Macmillan, New York

Bunce, R.G.H. and Shaw, M.W. (1973) 'A Standardized Procedure for Ecological Survey', J. Environmental Management, pp. 239-258

Burkholder, P.R. (1952) 'Cooperation and Conflict Amongst Primitive Organisms'. Amer.Sci., 40, 601-631

Burnett, J.H. (1964) Vegetation of Scotland, Oliver and Boyd, Edinburgh

Burrell, T.S. (1973) 'National Parks, the Big Three

- Conservation, Recreation and Education', J. Environmental Management, 1, pp. 201-205

Bush, R. (1973) The National Parks of England and Wales, Dent, London

CCN, (1986a) 'Conflict of Interests', Countryside Comm. News, 21, 1

CCN (1986b) 'A Promising Start', Countryside Comm. News, 23, 2

Cailiet, G., Stezer, P. and Love, M. (1971) Everyman's Guide to Ecological Living, MacMillan, New York

Caldwell, L.K. (1983) 'The Environmental Impact Statement: A Misused Tool', in R.K.Jain & B.L. Hutchings. (eds.), Environmental Impact Assessment: Emerging Issues in Planning: Urbana, University of Illinois Press. pp.11-25

Carrington, R. (1967) Great National Parks of the World, Weidenfeld & Nicolson, U.K.

Carson, R. (1962) Silent Spring, Hamilton, London

Cheremisinoff, P.E. & Morresi, A.C. (1977) Environmental Assessment and Impact Statement Handbook. Ann Arbor Science Publishers Inc., Michigan

Chorley R.J. & Kennedy, B. (1971) Physical Geography: A Systems Approach. Prentice Hall International, London

Christensen, E.O. (1955) Primitive Art, Bonanza Books, New York

Ciriacy-Wantrup, S.V. (1952) Resource Conservation, Economics and Policies, University of California Press, California

Clapham, W.B. (1973) Natural Ecosystems, The Macmillan Co., New York

Clark, B.D.; Chapman, K.; Bisset, R. & Walthern, P. (1976) Assessment of Major Industrial Applications. Dept. of the Environment, Research Report No 13, HMSO, London

Clark, B.D., Chapman, K., Bisset, R. and Wathern, P. (1978) 'Methods of Environmental Impact Analysis', Built Environment No.4 pp. 111-121

Clark, J.D. (1959) Prehistory of Southern Africa, Penguin Books, Harmondsworth

Clarke, G.L. (1965) Elements of Ecology, Wiley, New York

Clayton, K. (1971) 'Reality in Conservation', Geographical Magazine No. 44 pp. 83-84

Clements, F.E. (1916) An Analysis of the Development of Vegetation, Carnegie Institute. Publication No.242, Washington

Coddington, A. (1974) 'The Economics of Conservation' in A. Warren & F.B.Goldsmith, Conservation in Practice, Wiley, London, pp. 453-464

Collinson, A.S. (1977) Introduction to World Vegetation, George Allen and Unwin, London

Commoner, B. (1969) 'Evaluating the Biosphere', Science Journal 5A No.4, pp. 67-72

Commoner, B. (1972)The Closing Circle : Confronting the Environmental Crisis, Cape, London

Conant, F. et al (1983) Resource Inventory & Baseline Study Methods for Developing Countries. American Assoc. for the Adv. Sci., Washington, DC

Coppock, J.T. and Best, R.H. (1962) The Changing Use of Land in Britain, Faber, London

Costantino, I. (1972) World National Parks ; Progress and Opportunities, ed. J-J. Harroy, Hayez, Brussels, pp. 81-85

Cotgrove, S. (1976) 'Environmentalism and Utopia', The Sociological Review, 24, 23-42

Cotgrove, S. (1982) Catastrophe or Cornucopia, Wiley, Chichester

Cottrell, Sir A. (1978) Environmental Economics, Edward Arnold, London

Council on Environmental Quality (1973) Environmental Quality : Fourth Annual Report, Washington DC. Govt. Printing Office

Council on Environmental Quality (1975) Environmental Quality : Sixth Annual Report, Washington DC. Govt. Printing Office

Countryside Commission (1986) 'Conflict of Interests', Countryside Commission News, No.21 p.1

Cox, C.B., Healey I.N. and Moore, P.D. (1976) Biogeography. An Ecological and Evolutionary Approach, Blackwell Scientific Publications, Oxford

Crofts, R.S. & Cooke, R.U. (1974) Landscape Evaluation: A Comparison of Techniques, Univ. College London, Dept of Geography, London

Crosland, A. (1971) A Social Democratic Britain. Fabian Society, London

Cruikshank, J. (1972) Soil Geography, David & Charles, Newton Abbot

Cumberland, K.B. & Whitelaw, J.S. (1970) New Zealand, Longman, London

Curry-Lindahl, K. (1972) 'Africa : National Parks, Habitats Biomes and Ecosystems', World National Parks : Progress and Opportunities, pp. 105-118, ed. J.P. Harroy, Hayes, Brussels

Curtis, R. and Hogan E. (1980) Nuclear Lessons. An Examination of Nuclear Power's Safety, Economic and Political Record. Turnstone Press, Wellingborough, Northants

de Wit, C.T., Brouwer, R. and Penning de Vries F.W.T.(1971) 'A Dynamic Model of Plant and Crop Growth', Potential Crop Production, eds. Waring, P.F. and Cooper, J.P.,Heinemann, London, pp. 116-142

di Castri, F. and Mooney H.A. (eds) (1973) Mediterranean Type Ecosystems, Origin and Structures, Chapman and Hall, London

Daly, H.E. (1977) Steady State Economics, Freeman, San Francisco

Bibliography

Danserau, P. (1966) 'Ecological Impact and Human Ecology', Future Environments of North America, F.F. Darlind and Milton, J.P., (eds), Natural History Press, pp. 425-462

Darlington, C.D. (1969) The Evolution of Man and Society, Simon and Schuster, New York

Dasmann, R.F. (1976) Environmental Conservation, Wiley, New York

Dasmann, R.F., Milton, J.P., & Freeman, P.H. (1973) Ecological Principles for Economic Development, Wiley, London

Davidson, D.A. & Jones, G.E. (1986) 'A Land Resource Information System (LRIS) for Land Use Planning', Applied Geography. 6, pp. 255-265

Davidson, J. (1976) 'Discovery' in I.Wards, (ed.), New Zealand Atlas, A.R.Shearer, Wellington

Deevey, E.S. (1960) 'The Human Population', Sci. Amer., 203(3) 194-204

Detwyler, T.R. (1971) Man's Impact on Environment, McGraw Hill, New York

Dimblebey, G.W. (1967) Plants and Archeology, Baker, London

Dower Report. (1945) National Parks in England and Wales, Cmnd No. 6628, Ministry of Town and Country Planning, H.M.S.O. London

Duckham, A.N. & Masefield, G.B. (1970) Farming Systems of the World, Chatto & Windus, London

Duffey, E. (1970) Conservation of Nature, Collins International Library, London

Durand, J.D. (1971) 'The Modern Expansion of World Population', in T.R.Detwyler, (ed.), Man's Impact on Environment, McGraw Hill, New York pp. 36-49

Duvigneaud, P., Kestemont, P. & Ambrose, P. (1971) 'Productivite primaire des forets temperes d'essences feuillues caducifolies en Europe occidentale', in Duvigneaud, P. (ed.), Productivity of Forest Ecosystems. UNESCO, Paris, pp. 259-270

Eagles, P.F.J. (1984) The Planning and Management of Environmentally Sensitive Areas, Longman, London

The Ecologist, (1972) 'A Blueprint for Survival', The Ecologist, reprinted Penguin Books (1973), Aylsbury

Edmonson, W.T. (1971) 'Fresh Water Pollution', in W.W. Murdoch (ed.), Environment, Sinauer Associates, Inc. Stamford, Conn. pp 213-239

Egar, G. & Egar, J. (1979) 'Report of the New Zealand Recreational River Survey', 3 vols. National Water & Soil Conservation Organisation, Wellington, New Zealand

Egli, E. (1978) Switzerland. A Survey of its Land and People. Paul Haupte, Berne

Ehrenfeld, D.W. (1970) Biological Conservation, Holt, Rinehart & Winston, New York

Ehrlich, P.R., Ehrlich, A.H. & Holdren J.P. (1977) Ecoscience: Population, Resources, Environment, Freeman, San Francisco

Ehrlich, P.R. & Ehrlich, A.H. (1982) Extinction: The Causes and Consequences of the Disappearance of Species, Victor Gollancz, London

Ellington, A. & Burke, T. (1981) Europe: Environment, Ecobooks, London

Elton, C. (1958) Ecology and Invasion by Animals and Plants, Methuen, London

Elton, C. (1966) The Pattern of Animal Communities, Chapman & Hall, London

Environmental Resources Ltd. (1986) Resource Recovery: A Report for the Royal Commission on Environmental Pollution, HMSO, London

Evans, G.L. (1963) Environmental Control of Plant Growth, CSIRO, Canberra

Everhart, W.C. (1972) The National Park Service, Praeger, New York

Eyre, S.R. (1971) Vegetation and Soils: A World Picture, Arnold, London

Bibliography

Falla, R.A. (1976) 'Fauna' in I.Wards (ed.), New Zealand Atlas, A.R.Shearer, Wellington, pp.114-121

Farnsworth, N.R. (1982) The Consequences of Plant Extinction on the Current and Future Availability of Drugs, Univ. Illinois Medical Centre, Chicago

Fernie, J. & Pitkethly, A.S. (1985) Resources, Environment and Policy. Harper & Row, New York

Ferrar, A.A. & Kruger, F.J. (1983) 'South African Programme for the SCOPE Project on the Ecology of Biological Invasions', S.Afr.Nat.Sci.Prog.Rep., 72

Fischer, D.W. & Davies, G.S. (1973) 'An Approach to Assessing Environmental Impacts', J.Envir. Manag., 1, 207-227

Fisher, J., Simon, N. & J. Vincent (1969) Wildlife in Danger, Collins, London

Fitter, R. (1982) 'Wildlife', in W. Pettigrew (ed.), Conservation, Hodder & Stoughton, pp. 75-84

Flannery, K.V. (1969) 'Origins and Ecological Effects of Early Domestication in Iran and the Near East', in P.J.Ucko & G.W. Dimbleby, The Domestication and Exploitation of Plants and Animals, Duckworth, London

Flenley, J. (1979) The Equatorial Rainforest: A Geological History. Butterworths, London.

Folsom, C.E. (1979) The Origin of Life: A Warm Little Pond, Freeman & Co., San Francisco

Food & Agriculture Organisation (1982) 'Tropical Forest Resource', Forestry Paper, No 30, FAO, Rome

Food & Agriculture Organisation, FAO, (1983) 'Case Studies and Working Papers Presented at the Expert Consultation on Strategies for Fisheries Development, Rome, May 10-14, 1983, FAO Fisheries Report, No 295(Supplement), HMSO, London

Ford, A.B. (1970) 'Casualties of our Time', Science, 167, 256-263

Fothergill, S. & Vincent, J. (1985) The State of the Nation. An Atlas of Britain in the

Eighties, Pan Books, London

Frankel, O.H. & M.E.Soule (eds.), (1982) Conservation and Evolution, C.U.P. Cambs

Fraser-Darling, F.F. (1967) 'A Wider Environment of Ecology and Conservation', Daedalus,96, pp. 1003-1019

Freeman, A.M., Haveman, R.H. & Kneese, A.V. (1973) The Economics of Environmental Policy, Wiley, New York

Fuggle, R.F. (1983) 'Nature and Ethics of Environmental Concern' in R.F.Fuggle & M.A. Rabie (eds.), Environmental Concerns in South Africa, Juta, Capetown

Furley, P.A. & W.N. Newey (1983) Geography of the Biosphere, Butterworths, London

Gehlbach, F.R. (1975) 'Investigation, Evaluation and Priority Ranking of Natural Areas', Biological Conservation, 8, 79-88

Gilbert, L.E. & Raven, P.H. (eds.) (1975) Coevolution of Animals and Plants, Uni. Texas Press, Austin

Gilg, A.W. (1978) Countryside Planning, David & Charles, Newton Abbot

Gilg, A.W. (1981) 'Planning for Nature Conservation: A Struggle for Survival and Political Respectability', in R.Kain (ed.), Planning for Conservation, Mansell, London, pp.97-116

Gleason, H.A. (1922) 'On the Relation Between Species and Area', Ecology, 3, 158-162

Godley, E.J. (1975) 'Flora and Vegetation', in G. Kuschel (ed.), Biogeography and Ecology in New Zealand, Junk, The Hague, pp. 177-230

Goldsmith, F.B. (1974) 'An Assessment of the Nature Conservation Value of Majorca', Biol. Conserv., 6(2), 79-83

Goodland, R.J.A. & Irwin, H.S. (1975) Amazon Jungle; Green Hell to Red Desert, Elsevier, Amsterdam

Goudie, A. (1984) The Nature of the Environment, Basil Blackwell Publishers Ltd., Oxford

Green, B. (1985) Countryside Conservation, 2nd. ed. George Allen & Unwin, London

Guggisberg, C.A.W (1970) Man and Wildlife, Evans Brothers, London

Hall, A.V., de Winter, M., de Winter, B. & van Ousterhout, S.A.M. (1980) 'Threatened Plants of southern Africa', S.Afr.Nat.Sci.Progr.Rep. 45

Hall, D.O. & Nair, P.K.P. (1980) Agroforestry Species: A Crop Sheets Manual, International Council For Research in Agroforestry

Hansen, A.H. (1953) A Guide to Keynes, McGraw-Hill, New York

Hardin, G. (1968) 'The Tragedy of the Commons', Science, 162, 1243-1248

Harlan, J.R. (1967) 'The Plants and Animals that Nourish Man', Scientific American, 235, 89-97

Harris, D.R. (1972) 'New Light on Plant Domestication and the Origins of Agriculture: A Review', in R.L.Smith, The Ecology of Man: An Ecosystem Approach, Harper & Row, New York, pp. 73-87

Harris, L. (1984) The Fragmented Forest: Island Biogeography Theory and the Preservation of Biotic Diversity, Uni. Chicago Press, Chicago

Harris, M. (1971) Culture Man and Nature, Crowell, New York

Harrison, J.L. (1962) 'The Distribution of Feeding Habits Among Animals in a Tropical Rain Forest', J. Anim. Ecol., 31, 53-64

Haspel, A.E. & Johnson, F.R. (1982) 'Substitutability, Reversibility and the Development-Conservation Quandary', J.Envir.Manag., 15(1), 79-92

Helliwell, D.R. (1973) 'Priorities and Values in Nature Conservation', J.Envir.Manag. 1, 85-127

Hendee, J.C. (1978) 'Wilderness Management', Forest Service, U.S. Department of Agriculture, Miscellaneous Publication No 1365, Washington, D.C.

Hepting, G.H. (1964) 'Damage to Forests from Air Pollutants', J. Forestry, 62, 630-646

Heptner, V. (1972) 'The USSR and the Socialist Republics of Europe', in J.P. Harroy, World National Parks, Progress and Opportunities, Hayez, Brussels, pp. 127-131

Hibbard, W.R. (1972) 'Mineral Resources: Challenge or Threat?' in R.L.Smith (ed.), The Ecology of Man: An Ecosystem Approach, Harper & Row, New York, pp. 349-357

Hickey, R.J. (1971) 'Air Pollution', in W.W. Murdoch, Environment, Sinauer Associates, Stamford, Conn. pp. 189-212

Highton, R.B. (1974) 'Health Risks in Water Conservation Schemes', in L.C.Vogel et al, (eds.), Health and Disease in Kenya, E.Africa Literature Bureau, Nairobi, pp.175-178

Hill, R.D. (1982) 'Controlling the Epidemic of Hazardous Chemicals and Waste', Ambio, 12, 86-90

Hilton, K.J. (1971) 'The Lower Swansea Valley Project. The Study', in W.G.V.Balchin, Swansea and its Region, Handbook for British Association for the Advancement of Science, Uni. Coll. Swansea, pp. 365-370

Hobhouse Committee (1947) Report of the National Parks Committee, (England & Wales), Command No. 7121 H.M.S.O., London

Hodges, L. (1977) Environmental Pollution, 2nd. ed., Holt, Rinehart & Winston, New York

Hodson, H.V. (1972) The Diseconomies of Growth, Pan/Ballantine Books, London

Holdgate, M.W. (1979) A Perspective of Environmental Pollution, C.U.P. Cambridge

Horn, H. (1981) 'Succession' in R.M.May (ed.), Theoretical Ecology, Principles and Applications, Blackwell Sci. Pub., Oxford, Ch. 11

House, P.W. & Williams, E.R. (1977) Planning and Conservation. The Emergence of the Frugal Society, Praeger Publishers, New York

Howe, G.M. (1972) Man, Environment and Disease in Britain, David & Charles, Newton Abbot

Huntley, B.J. (1978) 'Ecosystem Conservation in southern Africa', in M.J.A.Werger, Biogeography and Ecology of Southern Africa, Junk, The Hague, pp. 1333-1384

Hutchinson, T.C. & Havas, M. (1980) Effects of Acid Precipitation on Terrestrial Ecosystems, Plenum Publishing Corp., New York

Hylander, C.J & Stanley, O.B. (1941) Plants and Man, The Blakiston Co., Philadelphia

Inglehart, R. (1977) The Silent Revolution. Changing Values and Political Styles Among Western Publics, Princeton, New Jersey

International Conference on Population (1984) 'Population, Resources, Environment and Development', Proceedings of the Expert Group on Population, Resources, Environment and Development, Dept. of Int., Economic & Social Affairs, Population Studies No. 90, United Nations, New York

Isakov, Yu. A. (1978) 'Scientific Bases of the Preservation of Natural Ecosystems in Zapovedniks', in J.G. Nelson et al. (eds.), International Experience with National Parks and Related Reserves, Dept. Geog., Faculty of Envir.Studies, Univ. Waterloo, pp. 527-548

IUCN (1980) World Conservation Strategy, International Union for Conservation of Nature, Gland, Switzerland

Ives, J.D. & Barry, R.G. (1974) Arctic and and Alpine Environments, Methuen, London

Iyengar, S.S. (ed.), (1984) Computer Modelling of Complex Biological Systems, CRC Press Inc.

Jacobi, R.M., Tallis, J.H. & Mellars, P.A. (1976) 'The Southern Pennine Mesolithic and the Ecological Record', J.of Arch.Sci., 3, pp. 307-320

Jarman, M.R., (1972) 'European Deer Economies and the Advent of the Neolithic', Papers in Economic History, ed. E.S. Higgs, C.U.P., London, pp.125-145

Jeffers, J.N.R. (1973) 'Systems Modelling and Analysis in Resource Management', Journal of Environmental Management, 1, pp. 13-28

Jones, G.E. (1979) Vegetation Productivity, Longmans, London

Jones, G.E. (1981) 'New Zealand's Scenic Rivers, a Study in Resource Use', Geography, 66.2, pp. 95-103

Jones, G.E. (1986) 'A Summary of Different Land Information Systems', Environmental Conservation & Development. Planning Exchange Occasional Paper, pp. 160-167

Kahn, H., Brown, W. & Martel, L. (1976) The Next 200 Years - A Scenario for America and the World, Morrow, New York

Kain, R. (1981) Planning for Conservation, Mansell, London

Keeton, W.T. (1980) Biological Science, 3rd. ed. W.W. Norton & Co., New York

Kellman, M. (1974) Plant Geography, Methuen London

Keyfitz, N. (1971) 'The Numbers and Distribution of Mankind', Environment : Resources, Pollution and Society, pp. 31-52, W.W. Murdoch, Sinauer Associates Inc., Stanford, Connecticut

Kleingebiel, A.A. & Montgomery, P.H. (1961) 'Land Capability Classification', Soil Conservation Service, Agricultural Handbook, No. 20, US. Dept. of Agric., Washington DC

Klinge, H. (1975) 'Biomass and Structure in a Central Amazon Rainforest', Tropical Ecological Systems, pp. 115-122, ed. F.B.Golley, Springer - Verlag, Berlin

Kormondy, E.J. (1969) Concepts of Ecology, Prentice Hall, Englewood Cliffs, New Jersey

Krebs, C.J. (1972) Ecology: The Experimental

Analysis of Distribution and Abundance. Harper & Row, New York

Kuschel, G. ed. (1975) Biogeography and Ecology in New Zealand. Junk, The Hague

Lack, D. (1946) The Life of the Robin, Collins, London

Leaky, L.S.B., Tobias, P.V., & Napier, J.R. (1964) 'A New Species of the Genus Homo from Olduvai Gorge', Nature, 202, 7-9

Leaky, R.E. (1982) The Making of Mankind, Abacus, London

Lee, N. (1982) 'The Future Development of Environmental Impact Analysis', J.Env.Manag., 14, pp. 71-90

Lee, R.B. (1968) 'What Hunters do for a Living, or How to Make Out on Scarce Resources', in Lee, R.B. & deVore, I., Man the Hunter, Aldine Press Chicago

Leopold, A. (1949) A Sand County Almanac, Oxford University Press, London

Leopold, L., Clarke, F.E., Hanshaw, B.B. & Balsley, J.R. (1970) 'A Procedure for Evaluating Environmental Impact', US.Geol.Surv. Circ., 645, US Geol.Survey, Washington DC

Levy, E.M. (1984) 'Oil Pollution in the World's Oceans', Ambio, 13(4), pp. 226-235

Lewis, M. & Clark, W. (1966) 'Journal of Exploratory Traverse of the Lewis and Clark Expedition 1804-1806', Field Study of Amer.Geog., pp. 13-26, Chicago University, Illinois

Lockie, J.D. et al, (1969) 'Breeding Success and Organochloride Residues in Golden Eagles in W. Scotland', J.App.Ecol, 6(3), 381-389

Lorrain-Smith, R. (1982) 'The Nature of Environmental Management', Journal of Environmental Management, 14, pp. 229-236

Lowe, P. & Goyder, J. (1983) Environmental Groups in Politics, George Allen & Unwin, London

Lucas, G. & Synge, H. (1978) The IUCN Plant Data Book, IUCN, International Union for Conservation of Nature and Natural Resources, Gland, Switzerland

Lucas, P.H.C. (1972) 'Australia and Oceania', in J.P. Harroy, (ed.), World National Parks Progress and Opportunities, Hayez, Brussels, pp. 97-104

McCaul, J, & Crossland, J. (1974) Water Pollution, Harcourt Brace & Jovanovich. Inc., New York

MacEwen, M. & MacEwen, A. National Parks, George Allen & Unwin, London

McHarg, I.L. (1969) Design with Nature, Natural History Press, Garden City, New York

MacKinnon, J.A. (1972) The Behaviour and Ecology of the Orang Utan, D. Phil. Thesis, University of Oxford

Mabberly, D.J. (1983) Tropical Rainforest Ecology, Blackie, London

Malone, T.F. (1976) 'The Role of Scientists in Achieving a Better Environment', Environmental Conservation, 3, pp. 83-89

Malthus, T.R. (1970) An Essay on the Principle of Population (1798), Penguin Books, Hamondsworth, Middlesex

Manabe, S. & Wetherall, R. (1975) 'The Effects of Doubling the CO_2 Concentration on the Climate of a General Circulation Model'. J. Atmos.Sci., 32, 3-15

Marsh, A. (1947) Air Pollution and Plant Life, Faber & Faber, London

Marsh, G.P. (1874) The Earth as Modified by Human Actions, Sampson Low, Marston Low & Seal, London

Marshall, J., Kinsky, F.C. & Robertson, C.J.R. (1977) The Fiat Book of Uncommon Birds in New Zealand, Vol.3, A.H. & A.W. Reed, Wellington

Martin, P.S. (1967) 'Prehistoric Overkill', Pleistocene Extinctions : The Search for a Cause, ed. P.S. Martin & H.E. Wright, Yale

University Press, Newhaven

Mason, H.L. & Langenheim, J.H. (1957) 'Language analysis and the concept of the environment'. Ecology, 38, 325-340

Massachusetts Institute of Technology (1970) Study of Critical Environmental Problems, MIT, Massachusetts

May, R.M. (1974) Stability and Complexity in Model Ecosystems, Oxford University Press, Oxford

Meadows, D.H.; Meadows, D.L.; Randers, J. & Beherns, W.W. (1972) The Limits to Growth, Earth Island Press, London

Meadows, M.E. (1985) Biogeography and Ecosystems of South Africa, South African Geography and Environmental Studies Series, Durban

Mellanby, K. (1967) Pesticides and Pollution, Collins New Naturalist Series, London

Mellanby, K. & Perring, F.H. (1977) Ecological Effects of Pesticides, Academic Press, New York

Mellars, P. (1976) 'Fire Ecology, Animal Populations and Man: a Study of Some Ecological Relationships in Prehistory', Proc. of the Prehist. Soc, 42, 15-45

Miller, A. (1961) Climatology, Methuen, London

Miller, G.T. & Armstrong, P. (1982) Living in the Environment, Wadsworth International Group Belmont, Calif.

Minckley, W.L. & Deacon, J.E. (1968) 'South-western Fishes and the Enigma of Endangered Species', Science, 159, 1424

Ministry of Housing and Local Government (1963) New Life for Dead Lands: Derelict Areas Reclaimed, HMSO, London

Ministry of Town and Country Planning (1947) Conservation of Nature in England and Wales, Report of the Wildlife Conservation Special Committee (England & Wales), Cmnd. 7122, HMSO London

Ministry of Town and Country Planning (1947) National Parks and the Conservation of Nature in Scotland, Cmnd. 7235, HMSO, London

Mitchell, B. (1983) Frozen Stakes: the Future of Antarctic Minerals, Int. Inst. Envir. Devel., N.Y.

Moncrief, L.W. (1970) 'The Cultural Basis for our Environmental Crisis', Science, 170, 508-512

Moore, N.W. (1968) 'Experience with Pesticides and the Theory of Conservation', Biol. Cons., 1, 201-207

Moran, J.M., Morgan, M.D. & Wiersma, J.H. (1980) Introduction to Environmental Science, W.H.Freeman & Co., San Francisco.

Morgan, R.K. (1983) 'The Evolution of Environmental Impact Assessment in New Zealand', J.Envir.Manag., 16, 139-152

Morgan, W.T.W. (ed.), (1973) East Africa its Peoples and Resources, O.U.P., Nairobi

Morton-Boyd, J. (1972) 'Conservation and the Scottish Landscape', Scottish Geographical Society, Lecture, Strathclyde University

Mossman, A.S. (1974) Towards Conservation, Intertext Books

Mumford, L. (1962) The Transformations of Man, Collier, New York

Mumford, L. (1966) The Myth of the Machine. Technics and Human Development, Harcourt, Brace & World, New York

Munn, R.E. (1979) Environmental Impact Assessment (Scope 5), J.Wiley & Sons, Chichester, England

Munton, R.J.C. (1974) 'Agriculture and Conservation in Lowland Britain', in A. Warren & F.B.Goldsmith, Conservation Practice, Wiley, pp. 323-336

Murdoch, W.W. (1971) Environment. Resources, Pollution & Society, Sinauer, Stamford, Connecticut

Bibliography

Murdock, P. (1961) Our Primitive Contemporaries, Macmillan Group, New York

Myres, N. (1976) 'An Expanded Approach to the Problem of Disappearing Species', Science, 193, 198-202

Myres, N. (1983) A Wealth of Wild Species, Westview, New York

Myres, N. (1984) The Primary Source. Tropical Forests and Our Future, W.W.Norton & Co., New York

Myres, N. (1985) The Gaia Atlas of Planet Management, Pan Books, London

Nash, R. (1967) Wilderness and the American Mind, Yale University Press, New Haven

National Geographic (1977) 'Preserving the Nation's Wild Rivers', National Geographic, 152.1, pp.2-59

Nature Conservancy Council (1984) Nature Conservation in Great Britain, Summary of Objectives and Strategy, HMSO, London

Nelson, J.G., Needham, R.D. & Mann, D.L. (1978) International Experience with National Parks and Related Reserves, Department of Geography, Faculty of Environmental Studies, University of Waterloo

Nicholson, M. (1970) The Environmental Revolution, Hodder & Stoughton, London

Nicholson, M. (1971) 'Environment on Record', Geographical Magazine, XXXXIII, pp. 108-114

Nicholson, M. (1973) The Big Change : After the Environmental Revolution, McGraw Hill, New York

Norton, G.A & Walker, B.H. (1982) 'Applied Ecology: Towards a Positive Approach', J. Envir. Management, 14(4), pp. 309-324 and 325-342

Ojo, G.J.A. (1978) 'Nigerian National Parks and Related Reserves', in J.G.Nelson et al, (ed.), International Experience with National Parks and Related Reserves, Department of Geography, University of Waterloo, pp. 271-293

O'Riordan, T. (1976) Environmentalism, Pion, London

O'Riordan, T. (1979) 'Ecological Studies and Political Decisions', Environment and Planning, 11, pp. 805-813

O'Riordan, T. & Hey, R.D. eds. (1976) Environmental Impact Assessment, Saxon House, London

O'Riordan, T. & Turner, R.K. (eds.), (1983) An Annotated Reader in Environmental Planning and Management, Pergamon Press, Oxford

Outdoor Recreation Resources Review Commission (1962) Outdoor Recreation for America, ORRRC, Washington DC

Owen, D.F. (1973) Man's Environmental Predicament, O.U.P., Oxford

Packard, V. (1960) The Waste Makers, Penguin Books, Harmondsworth, Middlesex

Paludan, C.T.N. (1985) 'The Future of Rural Systems in Space and Other Planets', Management of Rural Resources: Problems and Policies, An International Conference, July 14-20th, 1985, Univ.of Guelph, Ontario

Papadakis, E. (1984) The Green Movement in West Germany, Croom Helm, London

Park, C.C. (1980) Ecology and Environmental Management. A Geographical Perspective, Dawson Westview Press, Folkstone, Kent

Parker, T.J. & Haswell, W.A. (1962) A Textbook of Zoology, Macmillan & Co, London

Patterson, C.B. (1978) 'Walk of a Lifetime. New Zealand's Milford Track', National Geographic, 153.1, pp. 117-129

Paulick, G.J. (1971) 'Anchovies, Birds and Fishermen in the Peru Current', in W.W. Murdoch (ed.), Environment: Resources, Pollution & Society, Sinauer Associates, Stamford, Conn., pp. 156-185

Penning-Rowsell, E.C. (1983) 'County Landscape Conservation Policies in England and Wales', J.Envir.Manag., 16, 211-228

Perring, F.H. & Walters, S.M. (eds.) (1976) Atlas of the British Flora, 2nd. ed. Wakefield, BSBI/EP, Publishing

Perring, F.H. & Farrell, L. (1977) British Red Data Books 1: Vascular Plants, Soc. for the Promotion of Nature Conservation, Nettleham, Lincs.

Perry, L.M. (1980) Medicinal Plants of East and South Asia, M.I.T. Press, Cambridge, Mass.

Pettigrew, W. (ed.) (1982) Conservation, Hodder & Stoughton, Sevenoaks, Kent

Pfeiffer, J.E. (1972) The Emergence of Man, Harper & Row, New York

Phillips, A. (1985) 'National Parks or "National Parks"?' National Parks Today, 11, p.2

Phillips, J. (1974) 'Effects of Fire in Forest and Savanna. Ecosystems of Sub-Saharan Africa' in T.T.Kozlowski (ed.), Fire and Ecosystems, Academic Press, New York, pp. 435-482

Phillipson, J. (1971) Ecological Energetics, Edward Arnold Publishers Ltd., London

Pilbeam, D. (1972) The Ascent of Man. An Introduction to Human Evolution, MacMillan Publishing Co., Inc., New York

Pimental, D. (1981) CRC Handbook of Pest Management in Agriculture, Vol. III CRC Press Inc., Florida

Pimental, D. & Hall, C.W. (1984) Food and Energy Resources, Academic Press Inc., New York

Pinchot, G. (1936) 'How Conservation Began in The United States', Agricultural History, 11, pp.255-263

Platt, J.R. (1966) The Step to Man, Wiley, New York

Polunin O. & Huxley, A. (1981) Flowers of the Mediterranean, Chatto & Windus, London

Poore, M.E.D. (1964) 'An Approach to the Rapid Description and Mapping of Biological Habitats. Sub-Commission on Conservation of Terrestrial Biological Communities of the International Biological Programme, London

Population Reference Bureau (1985) Population Data Sheet, 1984, Population Reference Bureau : Washington D.C.

Prance, G.T. (1978) 'The Origin and Evolution of the Amazon Flora', Interciencia, 3(4), pp. 207-221

Pratt, C.J. (1965) 'Chemical Fertilizers', Reprinted in Chemistry in the Environment, Scientific American Book, Freeman Books, San Francisco pp. 97-108

Press, M.; Ferguson, P. & Lee J. (1983) '200 Years of Acid Rain', Naturalist, 108, (967), 125-128

Price, O.W. (1911) The World We Live In, Small, Maynard Co., Boston

Pryde, P.R. (1972) Conservation in the Soviet Union, C.U.P., Cambridge

Rasmussen, D.I. (1941) 'Biotic Communities of the Kaibab Plateau', Ecological Monographs, 3, pp.229-275

Ratcliffe, D. ed. (1977) A Nature Conservation Review, two vols., C.U.P., Cambridge

Reed, C.A. (1982) 'Extinction of Mammalian Megafauna in the Old World Late Quatenary', Bioscience, 2, pp. 284-288

Reid, L. (1961) The Sociology of Nature, Penguin Books, Harmondsworth

Regenstein, L. (1983) 'The Toxics Boomerang', Environment, 25(10) pp.

Report of the Countryside Commission (1968) Annual Report, HMSO, London

Report of the Nuclear Energy Policy Study Group (1977) Nuclear Power Issues and Choices, Ballinger Pub. Co., Cambridge, Massachusetts

Richards, P.W. (1970) The Life of the Jungle, McGraw Hill, New York

Riley, D. & A. Young (1969) World Vegetation, C.U.P. Cambridge

Roberts, T.M., Darrall, N.M. & Lane, P. (1983) 'Effects of Gaseous Air Pollutants on Agriculture and Forestry in the United Kingdom', Advances in Applied Biol., Vol 9. Academic Press, New York

Rowantree, R.A., Heath, D.E. & Voiland, M. (1978) 'The United States National Park System', paper 5 in J.G.Nelson et al. (eds.), International Experience with National Parks and Related Resources, Dept. of Geography, Uni. of Waterloo, Ontario

Russel, E.J. (1968) The World of the Soil, Collins New Naturalist, London

Salmon, J.T. (1975) 'The Influence of Man on the Biota', in G. Kuschel, Biogeography and Ecolgy in New Zealand, Junk, The Hague, pp. 643-661

Sandbach, F. (1980) Environment Ideology and Policy, Blackwell, Oxford

Sandford Review (1974) Report of the National Park Policies Review Committee, D.O.E., HMSO, London

Sauer, C. (1952) 'Agricultural Origins and Dispersals', American Geographical Society Bowman Memorial Lectures, Series 2, New York

Schultes, R.E. (1980) 'The Amazonia as a Source of New Economic Plants', Economic Botany, 33(3) pp. 259-266

Schwanitz, D. (1967) The Origin of Cultivated Plants, Harvard U.P., Cambridge, Massachusetts

Scott Committee (1942) Report of the Committee on Land Utilization in Rural Areas, Ministry of Works and Planning Cmnd 6378, HMSO

Sears, P.B. (1956) 'The Process of Environmental Change by Man', Man's Role in Changing the Face of the Earth, ed. W.L. Thomas, University of Chicago Press, Chicago, pp. 471-484

Selman, P.H. (1981) Ecology and Planning. An Introductory Survey, G. Godwin Ltd. London

Shchelkunova, R.P. (1976) 'The Lichen Cover Change by the Human Activity at the North of the Yenisei Basin', Symp. of Geogr. Polar Countries, XXIII, International Geographical Congress, pp. 136-137, Leningrad, Int. G.U.

Shriner, D.S. (1982) Acid Deposition : Effects on Terrestrial Ecosystems, Oak Ridge National Lab., TN, Environmental Science Division

Silvestrov, S.I. (1971) 'Efforts to Combat the Processes of Erosion and Deflation of Agricultural Land', in Natural Resources of the Soviet Union, eds. Gerasimov, I.P. et al. Freeman, San Francisco

Simmons, I.G. (1974) The Ecology of Natural Resources, Arnold, London

Simmons, I.G. (1978) 'National Parks in England and Wales', International Experience with National Parks and Related Reserves, in J.G. Nelson et al. (eds.), Dept. of Geog., Univ of Waterloo, pp. 383-410

Simmons, I.G. (1979) Biogeography : Natural and Cultural, Arnold, London

Small, G. (1968) The Virtual Extinction of an Extraterritorial Pelagiz Resource - The Blue Whale, Ph.D. Dissertation, Columbia University

Smith, P. (1975) Politics of Physical Resources, Penguin Books, Harmondsworth, Middlesex

Smith, R.S. (1983) The Use of Land Classification in Resource Assessment and Rural Planning. Inst. Terr. Ecol., Grange-over-Sands

Solomon, M.E. (1971) Population Dynamics, Inst.of Biology, Studies in Biology No. 18, Arnold, London

Soule, M.E. & Wilcox, B.A. eds. (1980) Conservation Biology, Sinauer Assoc.Inc., Massachusetts

Spedding, C.R.W. (1975) The Biology of Agricultural Systems, Academic Press, London

Spellerberg, I.F. (1981) Ecological Evaluation for Conservation, Arnold Sudies in Biology No. 133,

London

Stamp, D. (1974) Nature Conservation in Britain, Collins, New Naturalist Series, 2nd ed, London

Statham, D. (1972) 'Natural Resources in the Uplands: Capability Analysis in the New York Moors', RTPI Journal, 58, pp. 468-477

Stern, A.C. (1968) Air Pollution, 5 Vols, Academic Press, New York

Stewart, O.C.(1956) 'Fire as the First Great Force Employed by Man', Man's Role on Changing the Face of the Earth, ed. W.L. Thomas, pp.115-133, Chicago Univ. Press, Chicago

Stoel, T.B. & Scherr, S.J. (1978) 'Experience with EIA in the United States', Built Environment, 4, pp. 94-100

Stoker, H.S. & Seager, S.L. (1976) Environmental Chemistry, 2nd. ed., Scott, Foresman, Glenview, Illinois

Strahler, A.H. & Strahler, A.N. (1974) Introduction to Environmental Science, Hamilton Publishing Co., Santa Barbara

Strong, A.L. (1972) The National Park System: Ways and Means, National Parks for the Future, The Conservation Foundation, Washington DC

Tans, W. (1974) 'Priority Ranking of Biotic Natural Areas', The Michegan Botanist, 13, pp. 31-39

Tansley, A.G. (1935) The use and abuse of vegetational concepts and terms. Ecology, 16, 284-307

Tansley, A.G. (1946) Introduction to Plant Ecology, Allen & Unwin, London

Taranik, J.V. (1985) 'Characteristics of the LANDSAT Multispectral Data System', in R.K.Holz, The Surveillant Science: Remote Sensing of the Environment, Wiley, New York, pp. 328-351

Teal, J.M. (1957) 'Community Metabolism in a Temperate Cold Spring', Ecol. Monog., 27, pp.283-302

Theobald, R. (1970) The Economics of Abundance, Pitman, New York

Thirlwall, G. (1978) 'EIA - Taking Stock', Built Environment, 4, pp. 87-93

Thornthwaite, G.T. (1954) An Introduction to Climate, McGraw Hill, New York

Tivy, J.(1977) Biogeography : A Study of Plants in the Ecosphere, Oliver & Boyd, Edinburgh

Tivy, J. & O'Hare, G. (1981) Human Impact on the Ecosystem, Oliver & Boyd, New York

Tobias, P.V. (1967) Olduvai Gorge, Vol. 2, C.U.P., Cambridge

Trevelyan, G.M. (1964) Illustrated English Social History, Vol 3, Pelican Books, Harmondsworth

Trewartha, G.T. (1969) A Geography of World Population: World Patterns, Wiley, London

Tubbs, C.R. (1974) 'Woodlands : Their History and Conservation', in A.Warren & F.B.Goldsmith (eds.), Conservation in Practice, pp. 131-144, Wiley, London

Udvardy, M.D.F. (1975) 'A Classification of the Biogeographical Provinces of the World',IUCN Occasional Paper, No. 18, Morges, Switzerland

United Nations (1972) Conference on the Human Environment, Stockholm 1972, United Nations, Paris

United Nations (1984) Crisis or Reform: Breaking the Barriers to Development, U.N., New York

US Commission on Population Growth (1972) Population and the American Future, Washington DC, US Government Printing Office, N.Y.

US Department of the Interior (1971) 'Impact of Surface Mining on Environment', in T.R. Detwyler, (ed.), Man's Impact on Environment, McGraw Hill, New York, pp.348-369

Vernberg F.J. & Vernberg, W.B. (1970) The Animal and the Environment. Holt Rinehart Winston Inc, New York

Bibliography

Vink, A.P. (1975) _Land Use in Advancing Agriculture_, Springer, Berlin

Von Bertalnffy, L. (1962) 'General Systems Theory: a Critical Review', _General Systems_, 7, 1-20

Waller, R.A. (1970) 'Environmental Quality, its Measurements and Control', _Regional Studies_, 4, pp. 177-191

Walter, H. (1973) _Vegetation of the Earth_, English Universities Press Ltd., London

Ward, B. & Dubos, R. (1972) _Only One Earth_, Andre Deutsh, London

Warren, A. & Goldsmith, F.B. (1974) 'An Introduction to Conservation in the Natural Environment', in A. Warren & F.B. Goldsmith (eds.), _Conservation in Practice_, J. Wiley & Sons, London, pp.1-12

Waterhouse, D.F. (1974) 'The Biological Control of Dung', _Scientific American_, 230 (4), pp. 100-109

Watts, D. (1971) _Principles of Biogeography_, McGraw Hill, New York

Wells, A.K. (1960) _Outline ofHistorical Geology_, Thomas Murby & Co., London

West, R.G. (1968) _Pleistocene Geology and Biology_, Longman, London

Westing, A.H. (1977) _Weapons of Mass Destruction and the Environment_, Taylor & Francis, London

Wetterberg, G.A. (1974) _The History and Status of South American National Parks and an Evaluation of Selected Management Options_, Ph.D. Dissertation, Univ. of Washington, International Publication No. 74-29, 525, Vol. XXXV, No. 7, 1975, Ann Arbor, Michigan

Wetterberg, G.A., _et al_. (1976) 'An Analysis of Nature Conservation Priorities in the Amazon', _Brazilian Institute for Forestry Development_, Brazilia, Brazil

Wetterberg, G.A. & Meganck (1978) 'Columbian National Parks and Related Reserves : Research Needs and Management', in J.G.Nelson, _et al_, (eds.), International Experience with National Parks and Related Reserves, Dept. of Geog., University of Waterloo, pp. 175-232

Wheeler, G.M. (1966) 'Report on Geographical Surveys West of the 100th. Meridian', Field Study of American Geography, pp. 27-34, Uni. Chicago, Illinois

White, I.D., Mottershead, D.N. & Harrison, S.J. (1984) Environmental Systems. An Introductory Text, George Allen & Unwin, London

White, L. (1967) 'The Historical Roots of our Ecologic Crisis', Science, 155 (3767) pp. 1203-07

Whitmore, T.C. (1983) Tropical Rain of the Forests of the Far East, 2nd. ed. Oxford University Press, London

Whittaker, R.H. (1953) 'A Consideration of the Climax Theory; the Climax as a Population Pattern', Ecol.Monogr., 23, pp. 41-78

Whittaker, R.H. & Likens, G.E. (1973) 'Primary Production: The Biosphere and Man', Human Ecology, 1, pp. 357-369

Whittle, T. (1975) The Plant Hunters, Picador Books, London

Williams, R. (1973) The Country and the City, OUP, Oxford

Willis, A.J. (1973) Introduction to Plant Ecology, Allen & Unwin, London

Wilson, A.G. (1981) Geography and the Environment. Systems Analytical Methods, J. Wiley & Sons, Chichester, England

Wintsch, S. (1986) 'Corridors of Life', The Geographical Magazine, March, p. 108

Woodwell, G.M. (1970) 'Effects of Pollution on the Structure and Physiology of Ecosystems', Science, 168, pp. 429-433

Woodwell, G.M., Craig, P.P. & Johnson, H.A. (1971) 'DDT in the Biosphere : Where Does it Go?', Science, 174, pp. 1101-1107

Bibliography

Wooldridge, S.W. & East, W.G. (1962) The Spirit and Purpose of Geography, Hutchinson University Library, London

World Conservation Strategy (1983) The Conservation and Development Programme for the UK , Kogan Paul, London

World Resources Institute. (1986) World Resources 1986, Basic Books Inc., New York

Ziswiler, J. (1971) Extinct and Vanishing Animals, Springer-Verlag, New York

Zube, E, Brush, R. & Fabos, O. (1976) Landscape Assessment, Wiley, New York

INDEX